SNUBBING POSTS

Snubbing Posts

An Informal History of the Black River Canal

By THOMAS C. O'DONNELL

Author of
"The Sapbush Run," "Ladder of Rickety Rungs,"
Editor of "A Garden for You," etc.

Introduction by Albert B. Corey
State Historian

NORTH COUNTRY BOOKS
Box 331 — Old Forge, N. Y. 13420
Box 86 — Lakemont, N. Y. 14857
1972

SNUBBING POSTS
COPYRIGHT, 1949, BY
THOMAS C. O'DONNELL

First Edition — 1949
Second Edition — 1972

ISBN 978-0-932052-11-7

Preface

In a very special sense "Snubbing Posts" has been from the start a kind of cooperative enterprise, participated in by the author and the scores of old canal people who contributed so freely of their memories.

In the reports of engineers and various of the State's canal officials and of legislative committees, the compilation of canal enactments, and other official documents, could be found the complete year to year account of the statistical and budgetary details of the canal's operations, together with full engineering data, first on the building of the "Little Ditch," and then on its administration and maintenance.

In addition to this there was available a great body of human experience—stories that could be told by men who built the boats and of the men and women who navigated the canal and tended the locks, by the men who fed into navigation a steady stream of goods for distant ports, and by other men, who, in spite of niggardly appropriations from Albany, somehow kept the levels and locks fit for boating through seven and eight months of the year.

Out of these two bodies of data came "Snubbing Posts." And if the book seems to subordinate statistics to folks it is because the story, almost from the beginning, took charge of the business. Inquiries had discovered an unexpected number of old canallers, whose memories carried back over a period of seventy years and more. A band whose ranks regrettably were thinning with each passing year. And if this latter fact at first spurred me to redoubled effort in collecting their stories, presently I needed only the profound pleasure which each interview brought me soon to let the story have its way in favor of men and women—people who were young when the canal was new, equally with those of later generations who, if they had fewer service stripes, yet were no less proud of their part in one of the most colorful undertakings in the history of New York State. My thanks to one and all for their help in preparing the book and for the pleasure, as first visits were soon followed by casual calls, of having come to know them, and of having achieved in each a friend.

If in acknowledging my obligations I single out one among the old canal people it is because he has given of his time so far beyond the call even of our friendship, and also because he may serve as a symbol for the scores of others. Referring to Mr. Roscoe Clark, who in 1879 began a canal career that ended only with the Canal's abandonment in 1922. His never failing memory of men and events, covering the entire reach of the canal from Rome to Carthage, has enabled me in more than one instance to resolve conflicting data. It is his happy privilege to live, not only by the side of a road of surpassing beauty, but to look across the road there upon the series of five combined locks over which, after a long boating experience, he presided during the last twenty-seven years of the canal's existence.

The research part of the project was made less arduous by friends who have given willingly of their time and counsel. Miss Helen Salzmann, Librarian of the Jervis Library Association in Rome, placed at my disposal the store of canal material contained in the library of the late John B. Jervis, and Miss Ruth Traxel, Librarian of the Erwin Public Library, Boonville, rendered untiring service in procuring invaluable State canal documents, with Mrs. Edith B. Swancott, Librarian of the Oneida Historical Society, organizing for me the volume of original documents found in the collections of the Oneida Historical Society.

For access to old newspaper files in the Black River country, and for many other courtesies, I am indebted to Mr. Robert C. Rich, Editor of the Carthage *Republican Tribune,* and to Mr. A. Karl Arthur, Editor of the Lowville *Journal and Republican.* To them, and to Mr. Clayton A. Musser and Mr. Alfred C. O'Donnell, Editors of the Boonville *Herald,* in which the "Snubbing Post" chapters originally appeared as articles, my deepest appreciation.

Mr. John A. Scott, President of the Rome Historical Society, out of his profound knowledge of the Rome, the terminal, end of the canal, placed at my disposal much historical lore growing out of the backgrounds of the Erie Canal in the Rome reaches.

And for his more than kind introduction to the book, I am under renewed obligations to Dr. Albert B. Corey, New York Historian, who, because he liked also the "Sapbush Run," older brother to "Snubbing Posts," strengthened my resolution in the first instance to undertake the story of the Black River Canal.

THOMAS C. O'DONNELL.

Contents

CHAPTER		PAGE
	Introduction	11
1.	How It Was	15
2.	Surveys and Surveys of Surveys	21
3.	The Feeder Opened to Navigation	29
4.	Boating Starts	36
5.	Whistle 'Round the Bend	43
6.	The Breaks	50
7.	Fun on the Canal	58
8.	Early Boats and Boatmen	65
9.	Family Business	70
10.	More About Boat Building	77
11.	Trials of Traffic	83
12.	The Breaks	89
13.	Timber Becomes Big Business	96
14.	Down-River Lumber	100
15.	Forestport and Thereabouts	107
16.	Lock Tenders	113
17.	More About Lock Tenders	120
18.	Still More Lock Tenders	127
19.	Disturbing News	134
20.	The Pulmotor Fails	140
21.	Dreams Don't Jell	147
22.	Curtains	154
	Postscript	159

List of Illustrations
(Between Pages 64 and 65)

The "Ollie," Sweetheart of the Feeder Fleet

The "George F. Weaver"

The Five Combines

State Shop in Boonville

Nelson J. Beach *(portrait)*

The steamer "Edith M. Van Amber"

Peter Philips *(portrait)*

Lock No. 71 in Boonville

The "Hiram and Wilber"

The "Dudley Capron"

Sluiceway at lock 35

The "Annie Laurie"

Remains of lock at Otter Creek

The "Timothy Curtin"

Lumber-laden boats at Otter Creek

Model of a timber "crib"

Looking north from lock 35

Model of a "laker"

View of canal and boat at Dunn Brook

The Henry Abbey saw mill on Independence Creek

Basselin mill at Castorland

Van Amber mill in town of New Bremen

Same, with boat-building plant

The Pasenger hotel at Bush's Landing

The Wellington Brown hotel in the town of Greig
The steamer "J. F. McCoy" at Lyons Falls
Repairing bridge at Glenfield
The "Grover Cleveland"
"The Harrison and Kenneth"
View between the "Lower Threes" and the Five Combines
The Otter Creek dam
Culvert No. 4 in the Lansing Kill valley
Baker's Falls at lock 70

Introduction

Nothing on the Black River canal was ever hurried. It took years to work up enough momentum in the Legislature to get it started and years more to get it built. It never was completed north to Ogdensburg as its promoters had hoped. It seemed to be chronically troubled with insufficient water in dry summers and too much when freshets broke down its banks in spring. Even when water broke through the banks and left boats high and dry there was no great speed in repairing the breaks. People knew what it was to take their time, go visiting, swap yarns, and relax. Journeys up and down the canal were unhurried. Tom O'Donnell has captured the spirit of the canal so that reading SNUBBING POSTS becomes an adventure in relaxation.

If you have not seen the canal, go see it at once. You will never see its like anywhere else. Follow it, especially from Boonville to Rome. Through this winding river valley the canal goes down hill, not down the gentle slope of a rolling plain but down the precipitous descent of a hilly ravine. Why, you will ask, did anyone build a canal this way, for it must have been obvious that travel would be obstructed by lock after lock, and that costs of maintenance would be high? The answer: a feeder for the Erie Canal and a highway for the produce of the North Country.

It lasted seventy years—"not a long time," as the author remarks, "measured against the centuries, but it was long enough for a great tradition to grow up around the canal; long enough for boys to be born of boating parents, and themselves in turn to be buried by sons who piloted their own boats." It came to an end almost imperceptibly in the 1920's when commerce simply stopped and the Legislature provided no more funds for its upkeep.

Some histories of canals are stuffy. Not this one. Tom O'Donnell calls it "an informal history," and in it you feel the spice of living all the time. I hope he never runs out of things to write about in the North Country.

ALBERT B. COREY,
State Historian.

Albany, N. Y.,
    August 15, 1949

SNUBBING POSTS

1

How It Was

It is not easy for a generation that has lost its sense of remoteness to comprehend the difficulties which the idea of a canal up to and through the Black River valley had to surmount before the statesmen at Albany could be forced into the position of having to authorize construction. The crew of an airplane today takes off on a hop to China with little more to-do than a farmer in Greig in those ancient times set out with a load of potatoes for Lowville. Today no spot on the globe is remote; in the first quarter of the last century the valley of the Black River was just an uncertain streak on the inaccurate maps of the day. For that matter, the famous Sauthier map, published in London as recently as 1777 had not so much as shown the Black River, while the Morse map of 1796 had indicated the river, but given it as flowing into the St. Lawrence River at a point nearly due north of Boonville.

It was in 1825 that Governor DeWitt Clinton first suggested the idea for a canal to be run from the Erie Canal up to and down through the Black River country.

By that year Lewis County could boast of 12,000 people, whose sense of remoteness had not deterred them from making the long journey from, in the great majority of cases, Connecticut. Once here, however, the sense of remoteness took hold in a kind of reverse form, for rare were the instances when they left the valley, so hazardous was the journey, by any route, down into the valley of the Mohawk. Cases are recorded in which a settler, with winter closing in, did bare himself to a trip back home, leaving his wife and children, the larder properly stocked, to the whims of wild beasts, and the even wilder weather, but always at their beck the never failing kindness of the nearest neighbor. Come spring he would return, bringing not only needed supplies but still other settlers.

Such roads as traversed the valley were unworthy of the name. Travel was still mostly by horseback, and a quarter of a century was yet to pass before a plank road would be run from Utica to Boonville. The celebrated "State Road" around 1810 entered the North Country environs at Prospect, in the town of Trenton, after having been completed thus far from Johnstown; thence it proceeded to a point on the present State road between Remsen and Alder Creek, ultimately reaching Watertown by way of Boonville and the upper road to Lowville, and so on to Watertown. Shortly after the State road came along, a series of corporations completed, section by section, a turnpike from Utica to Watertown, intersecting the State Road below Alder Creek, and beyond that point utilizing the State Road. But that was fifteen years prior to the Clinton project and it had fallen into a condition that made travel over it hazardous in the extreme, yet it was the only way out of the valley except by old roads that by now were little better than trails.

Thus it was that settlers, once in the valley, were there pretty much for keeps; for the same reasons, the world outside knew as little of how things went here as the settlers knew of goings on in New York or Boston.

Just how Clinton in the first place got the idea of running a canal up over the Oneida County hills and then letting it down again into the Black River must be a matter of conjecture. At any rate the suggestion does not seem to have staggered the rest of the State. By this time the public had become used to seeing his dreams made good—and, besides, the voters at the previous election had sent him back to Albany with a resounding majority that silenced his political enemies and amounted to a mandate to go on in the strenuous, Clintonian manner to which they had become accustomed.

Needless to say, the idea of having a canal in their midst delighted the people of the valley, what with the prospects of getting the produce of their soil to outside markets. Getting Black River goods to market, however, was far from being a major concern elsewhere. The Erie Canal, now completed, had given New York a virtual monopoly as a market for the produce of the great grain raising areas of the Middle West, and little thought was bestowed on a region so difficult of access as Lewis and the adjacent counties. Citizens of the section, and their representatives in the Legislature, however, were vocal enough in their clamor for a canal, once the idea was projected.

It was, as a matter of fact, eleven years later, in 1836, when agitation for enlarging the Erie Canal was at its height, that a lone voice from Albany put in a notable appeal for a canal into the Black River lands. It was the voice of a Senate committee appointed to report on a new deluge of North Country petitions that was flooding both branches of the Legislature. The report, presented by the committee's chairman, Levi Beardsley, of the then Sixth Senatorial District (Lewis, Jefferson, Herkimer, Madison and Oswego composed the Fifth District), put the economic aspects of the Black River valley case when it said:

"With our present canals, a merchant at Green Bay or Chicago can obtain his goods almost as cheap as a merchant at Lowville, and the agricultural products of the great West will find their way to the seaboard at about as low a price as from the section of the country intended to be benefited by this canal. The tendency of this policy is to keep the price of real estate stationary, if not to depress it, in the secluded counties, and not only to retard their settlement and improvement, but to draw off their inhabitants to more favored regions. Such has been the effect of our present canals, and such will continue to be the result to a greater extent as the Erie canal shall be enlarged, and thus diminish the expense of transportation."

By now, however, it had become obvious that the Rome level of the Erie, being the summit level, must, if enlargement became a fact, have a new source of water beyond small streams like the Oriskany and Fish Creeks and the Lansing Kill. Especially must a source of water be found that would supply an unfailing amount of water in the dry months of summer. This meant storage reservoirs. Such a supply was offered by the Black River headwaters regions, although Fish Creek at its headwaters had its advocates.

The engineers almost to a man favored a plan offered long before for a canal to be run to Boonville, with a feeder bringing water from the Black River at what later was to be known as Forestport, with reservoirs to be built in the wilderness beyond, notably at North Lake and Woodhull. A sentiment had persisted, however, in the Legislature that, not a canal, but a mere feeder should be run from Rome to the river at Forestport, should this source of water be selected.

The situation thus presented stirred the entire region, its contention neatly set forth by the committee report just referred to:

"Those asking for the Black River Canal will scarcely perceive the propriety of expending nearly $200,000 [the estimated cost arrived at by the engineers] for a feeder only (and that perhaps an inadequate one and where the water can scarcely be spared), and their section of country excluded from a participation in our canal system. The committee, therefore, believing that ultimately the Black River will have to be resorted to as a feeder for the Erie canal, are decidedly of opinion that whenever resorted to, a canal should be constructed."

The Legislature was rapidly making up its mind—so rapidly that when Beardsley reported a bill authorizing construction of the canal, the affirmative vote was seventeen to ten. The struggle was over, provided Assembly action was favorable, and it was. On November 11th of the next year, 1837, the first contracts were let for actual construction.

If these paragraphs have stressed the importance of the Black River valley, in relation to the entire reach of the canal from Rome to Carthage, it is because the results that followed the arrival of the canal were so dramatic in the release of pent-up power latent in a region so completely land-locked as was Lewis County. Almost overnight the forests along the east side of the river, converted into lumber, were being loaded onto canal boats and carried to markets in Troy, Albany and New York. And across the river wide areas of rich farming lands were in good time rivalling the forest industries in the volume of their yield.

The complete story of American transportation when written will have as one of its most glowing chapters a full account of the utter revolution in the life of that once remote region by a waterway that took goods from a place where they were produced and set them down in a place where ready money could be had for them.

Results quite as startling came at the same time in northern Oneida County to a tiny collection of shanties that made no pretense to being so much as a village, that had not dreamed of a place in the economic sun—that, indeed, was content to turn out a decent day's stint in the little mill alongside a gratifying water site on Black River. "Smith's Mill," the name was, at such rare times as it was called anything at all. Then one day surveyors came in and the mill people learned that a canal feeder, whatever a canal feeder was, had been thought of, to tap the river at that point and carry part of its water to Boonville, there to collect in a pool and be distributed southward as the needs of the Erie demanded, and northward as the traffic of the canal to and from Lyons Falls might call for. The feeder came; Smith's Mill became "Williamstown," and in good time Forestport. And overnight the place was the heart of a lumbering industry that rivalled, in magnitude and drama, the down-river operations. Their fenced-in condition had cost the denisons of Smith's Mill no sleepless nights, and there is no record that when the tornado hit they so much as batted an eye.

Every region traversed by the canal, for that matter, came into a new life. Boonville overnight found itself invested with high importance in North Country affairs. Its canal basin was lined with commercial houses having to do with the canal trade, manufacturing enterprises that found here brisk markets in a countryside to which the canal brought money in exchange for its produce, and boat-building plants that supplied the canal trade with some of its sturdiest craft.

The valley of the Lansing Kill felt its pulses quicken. Saw mills began to appear where none were before, while the business of keeping the multitudinous locks operating, and the boats moving in an even tempo, called to the effort pretty much the entire countryside—men who were to contribute to the canal saga more than their full share of color and legend.

In the original plan the canal was to leave the Black River at Carthage and proceed on north to Ogdensburg. Gradually, however, the St. Lawrence dropped out of the discussions, and when the plans ultimately had jelled the State decided to drop its tools at Carthage and call it a day. It was just as well, considering the later tribulations of keeping in a fair working condition even the shorter canal, and considering also that by the time the canal reached Carthage Ogdensburg had a railroad, the Great Northern, running to the east, while soon its citizens would be taking the cars to Watertown, and even to Rome and New York, should they feel themselves up to the trip.

Such were the backgrounds against which the idea of the Black River Canal was born, and after incessant agitation by a determined North Country public, and in spite of a lot of political hocus-pocus in Albany, came to be built and, built, served the public, in spite of niggardly appropriations for its maintenance, for more than seventy years.

Seventy years is not a long time, measured against the centuries, but it was long enough for a great tradition to grow up around the canal; long enough for boys to be born of boating parents, and themselves in turn to be buried by sons who piloted their own boats along the Lansing Kill reaches, and on over the Boonville hump to Lyons Falls, there to lock into the river and, under tow by the "L. R. Lyon" or other river steamers, snub up at Carthage. Lives woven of the stoutest stuff, but embroidered with silken threads of many hues and patterns, humor or tragedy the dominant note, or pathos, or a mixture of all three. Or it might be a story of just simple, every-day doing of the job according to one's lights. Of all such—of the men who ran the boats and tended the locks and kept the canal in repair and produced the goods that made up the cargoes, and, of course, the men who built the canal, and of how they did it—of all these will the chapters that follow this introductory one be made.

2

Surveys and Surveys of Surveys

The thirteen years that followed Governor Clinton's suggestion for a Black River Canal was just one long period of surveys, and re-surveys of surveys. North Country reaction to the Governor's pronouncement was so immediate as to bring authorization of a survey of two possible routes northward from the Erie. One survey would leave the Mohawk River from Herkimer, the other from Rome or thereabouts. James Geddes, who had emerged from Erie Canal construction with high rating as a top-flight engineer, was assigned to the task. Having completed the Herkimer route Geddes found his instructions vague as to whether the second survey should proceed from Rome northward by way of Boonville or of Camden. He solved the problem by making two surveys.

Geddes' report favored the Boonville over the Herkimer route, but he presented his data as between the Boonville and the Camden route and let it go at that. On the basis of lockage and costs Camden would have won out in the long run. The route from Rome through Boonville would be seventeen miles shorter than the Herkimer route, with a lockage advantage of 180 feet. As between Boonville and Camden the latter presented a lockage of only 635 as against 1587 feet by way of Boonville —all three surveys were for a canal to reach the St. Lawrence at Ogdensburg. The Camden route, however, by crossing the Black River in its lower reaches would by-pass upper Oneida and Lewis Counties, from which came the greatest clamors for a canal.

It is notable that Geddes overestimated the lockage by way of Boonville. He called for 198 locks, each with a lift of eight feet, while his estimates of total cost to Ogdensburg were pitifully low—$931,014. The real lockage of the canal as ultimately built was 1,082 feet, and the number of locks between Rome and the river at Lyons Falls 109.

The Legislature of 1826 regarded the Geddes report as lacking in important details and called upon the Canal Commissioners for another survey. A year passed without action of any kind, but in Lewis County there was action, and to spare, to take up the slack. Mass meetings were held and memorials to the Legislature were drawn up, and the lawmakers in 1827 were forced to consider some kind of action, only to decide that the entire business must await a turn for the better in the State's finances. Another flood of petitions suddenly began arriving in Albany, but these contained something new—they importuned the Legislature to authorize the formation of a stock company empowered to build a canal from Rome, by way of Boonville, to High (now Lyons) Falls, and from that point by a canallized river to Carthage. The Senate Committee appointed to consider the petitions reported favorably, presenting the case for the North Country more succinctly than did the petitions. In studying the claims of the petitioners, they were, they said, impressed with "one very important consideration in favor of the petitioners, which, if their views are right, presents strong claims to the Legislature in favor of their application.

"This river passes through a rich and fertile country, abounding in valuable timber and inexhaustible beds of bog and mountain ore, and in every respect well calculated to sustain a dense and flourishing population. Connect this river with the Erie Canal, and it would be equal to an artificial canal of one hundred miles, costing one million dollars, for the reason that the speed on it would be nearly double, and cost nothing except . . . one dam to keep it in repair." It was estimated that the cost of the entire canal would be but $400,000.

A bill of incorporation was made law and the commissioners of the new corporation (the "Black River Canal Company") itself ordered a new survey, for the job selecting another distinguished Erie Canal veteran, Alfred Cruger. Cruger in reporting on his survey recommended a canal to follow substantially the route that in after years was adopted. Had his recommendations been followed, however, the Lansing Kill gorge would have seen some picturesque adjuncts added to its natural scenery. Cruger, to overcome the sharper grades along the Kill, advised the use, in the place of locks, of "inclined planes," such as were in use on the Morris Canal in New Jersey. These contrivances operated on the same principle as afterwards were to be used in Cincinnati and other

American cities in hauling streetcars up the steep hills, and, of course, letting them down again. The operation and mechanisms were simple enough. A boat entering a lock would find itself on a kind of saddle, which would be hauled up the grade on a track soundly anchored in a base of rock. The motive power was supplied by an engine a-top the plane, turning a drum over which rolled ropes attached to the saddle. At the top the saddle moved into a lock, and from this point on, until it reached the next plane, the boat was on its own. There were two tracks, it should be added, which made it possible, when traffic was brisk, to let down a boat at the same time that another was drawn up.

One Morris Canal plane, eight hundred feet long, overcame a grade of eighty feet, which, with a lift of ten feet per lock, which was the rule with the Black River Canal as ultimately built, would have required eight locks. The plane, while it would have added a Coney Island note to the Lansing Kill prospect, had a strong point in its favor—speed. The lift at the place just referred to was made in fourteen minutes; eight locks would have required better than an hour. In the end, however, the plane lost out, and because of its inability to handle boats of the size required for North Country traffic. The New Jersey boat was only sixty feet long and eight and a half feet wide, as against boats up to ninety feet long and fourteen feet wide that would be accommodated on the Black River Canal. Cruger's opinion was supported by a check of his figures made by Benjamin Wright, Rome's great engineer, who knew the North Country intimately through extensive land surveys which he had made over a period of more than thirty years.

All the efforts of Cruger and Wright were cancelled out, however, when the public refused to buy stock in the project. The company's commissioners turned the project back to the State, upon which the Legislature, in 1831, after a year's delay, called for a new survey of the route to be made by Holmes Hutchinson, another engineer who came out of the Erie construction with a high reputation. Hutchinson's report, like those which had preceded it, advocated the use of planes along the Lansing Kill, and, like previous surveys, this one came to naught. Again petitions from the North Country asked the legislature for a stock company, to build either a canal or railroad into the valley—anything that would connect them with the Erie. The ever responsive Legislature in

1832 authorized such a company, to be incorporated with a capital stock of not to exceed $900,000, with shares at $50 each.

The commissioners of this new company, unwilling apparently to break a tradition, did nothing. And did it so thoroughly as not to have a new survey made. This went on until 1834, when again in response to a new flood of North Country petitions, the Legislature ordered the Canal Commissioners to make a new survey. This time the greatest of all Erie alumni, a man whose great abilities had already won him renown in the new field of railroad engineering and construction, was called to the task—Timothy B. Jervis, of Rome.

Jervis' contribution to the accumulating body of survey reports was his rejection of planes as a practical means of overcoming the steep Lansing Kill grades. He did compromise to the extent of including two planes in his plans, but stated that should they prove not practicable they would readily be replaced by locks.

The Jervis report did not rouse the commissioners to action and two more years passed. The end of the comedy, however, was in sight, for in 1836, as related in the previous chapter, the Legislature, goaded on by new agitation in the northern counties—Oneida, Lewis, Jefferson, and St. Lawrence—and the masterful appeal of the Senate Committee appointed to report upon a memorial presented by a new mass movement in the North Country, put the memorable Act chapter 157, Laws of 1836, upon the statute books. A further nudge was given the Legislature by a growing conviction that increased supplies of water must without delay be found for the Erie Canal when enlarged—and enlargement was already under way.

The memorial that played so important a part in this sudden denouement well deserves a place in the record. Two years previously (that would be 1834) the Legislature had asked the Canal Commissioners to make a survey of the resources of the region to be served by a canal. In their 1835 report the Commissioners declared that "from the multiplied and pressing duties of the acting Commissioners, it has not been practicable for them or either of them to engage in collecting the facts, much less to devote to the subject the time and attention necessary to correctness."

Consequently a mass meeting was held in Lowville on August 18th of that year—1835. It was not just another mass meeting. Plans for it had been carefully laid, and the program of action which it adopted was of blitzkrieg proportions, and holding a lesson for all communities to whose needs a legislative body turns the deaf ear. The gathering had delegations from localities all the way from Rome to the St. Lawrence. The chairman of the day was Eli West, of Wilna, and Henry A. Foster, of Rome, was secretary. Oratory was fiery that day, and when the last "I thank you" had been uttered action took over. First off, a committee was appointed to undertake a survey of possible tonnage for the Canal such as the Legislature had asked of the Commissioners. Charles Dayan, one of Lowville's first citizens, was a member, as was Nelson J. Beach, of Watson, and Henry Graves, of Boonville, Henry A. Foster, of Rome, and Patrick S. Stewart, of LeRay. The choice of Beach brought into the forefront of the struggle for a canal a fighting spirit that was never to let up until the job was completed in 1855, then taking up the fight against mismanagement, and too often a complete neglect in Albany of the interests of both the canal and the regions which it was built to serve. He was to serve, and with distinction, as Canal Commissioner, Canal Appraiser, Assemblyman, and Senator. A farmer, he had become a self-taught surveyor, and, entrusted with important surveys of forest regions lying to the east of the Black River, he had a practical knowledge of the North Country and its resources possessed by few men.

The memorial drawn up by the committee presented figures for the Oneida towns of Lee, Western, Steuben, Boonville, the northern part of Remsen, all of Lewis County, and those parts of Jefferson and St. Lawrence that would be more or less contiguous to the canal. The report arrived at the figure of 36,113 tons of goods brought into the region thus represented and shipped out during the previous year. The largest single item was for lumber, including shingles—6,258 tons. The memorial stressed the sudden stimulus which would be given to the lumber industry with the coming of the Canal, and declared that "not less than fifteen million feet of pine lumber, or 15,000 tons, will be ready for market upon the opening of the canal." This was for lumber; in addition there would be 10,000 shingles, or 2,000 tons; 200,000 cubic feet of square timbers in rafts, or 4,000 tons. The last-named item would never materialize, but when the committee turned a prophetic eye to the production and shipment of spar and pile timbers its figures did not begin to correspond, as will be seen in later chapters, with the vast proportions that it was to reach; the committee's figures were for 2,000 tons, or 100,000 cubic feet.

All these and other statistics, covering so wide a territory, were made possible by the fighting-mad spirit which prevailed at that August 16th meeting. An important part of the drive set up that day was a "corresponding committee," its duty—to ascertain "the probable amount of iron and lumber in their respective towns which would be transported on said canal, their estimates to be turned over to the memorial committee. The following were named as members: representing the Rome area, Jay Hatheway; Lee, Anson Dart; Western, Harvey Brayton; Boonville, Philip Schuyler; Remsen, Andrew Billings; Steuben, Merritt Brooks; Leyden, Thomas Baker; West Turin, Homer Collins; Turin, Henry Regan; Martinsburg, Daniel S. Baitey; Lowville, Sylvester Miller; Watson, Nelson J. Beach; Gregg, Caleb Lyon; Diana, Thomas Brayton, Jr.; Denmark, Harrison Blodgett; Pinckney, Jacob B. Yendez; Harrisburg, Elias Gallop; Champion, Alfred Lathrop; Rutland, John Felt; Watertown, George S. Sherman; LeRay, Oliver Child; Philadelphia, William K. Butterfield; Alexandria, Archibald Fisher; Antwerp, Charles B. Hoard; Wilna, Joseph C. Budd; Fowler, T. G. Fowler; Gouverneur, Edwin Dodge; Rossie, Solomon Pratt; Edwards, Hubbard Goodrich; Hammond, Sylvester Butrick, and DeKalb, Asa Sprague.

Some of these men were destined not to see the canal completed. The will and enthusiasm with which they applied themselves to this assignment, however, helped mightily to assure the canal's authorization, and for their contribution to the effort they deserve memorialization far beyond this passing call of the roll.

The meeting adjourned itself to October 1st. This new gathering was notable for the size of the turnout: from Oneida County, seventeen; Lewis, 146; Jefferson, sixty-six, and St. Lawrence six. Among the new figures that day were that great Roman, John Stryker, afterwards to write his name in railroad history in the Middle West; Henry Graves, of Boonville; Lemuel Hough, of Remsen, and General Ela Merriam, of Leyden.

In addition to adopting a set of resolutions, which little more than reiterated the arguments presented over the previous ten years, the meeting appointed a committee to draw up an address to the people of the State. The address was duly written and, together with the proceedings of the two meetings, was presented with the memorial to the Legislature. Next year that very deliberative body saw the light. This could go on and on, the members must have reasoned; the North Country dander is up and capitulation, since it must come some time, might as well come now.

Immediately upon the passage of the authorization act by the Legislature, the Canal Commissioners detailed still another Erie veteran, Portous R. Root, to inspect the operation of the Morris Canal in order to settle once and for all the claims of inclined planes as against locks. He returned from New Jersey with a veto of all planes, in consideration of the necessity of using larger boats than those plying the New Jersey canal. Larger boats meant a larger prism, one to conform to the dimensions of all other canals lateral to the Erie. Accordingly, a canal forty feet wide at the surface, twenty-six at the bottom, and carrying four feet of water, was decided upon as Jervis had recommended. The dimensions of the canal as it was actually built, however, were forty-two, twenty-six and four feet, and with the Forestport feeder having dimensions of forty-six, thirty and four feet.

Root, who had been appointed chief engineer of the canal, gave himself to the first step leading to actual construction—a location survey. This he completed in 1837. Contracts were soon let, and construction started in 1838 between Rome and the mouth of the Lansing Kill.

With only the slightest deviation construction followed the location survey. Between Rome and Westernville two routes had been under discussion. One, from the latter place would follow the general direction of the Mohawk into and through the village of Delta direct to Rome; the other would follow the eastern side of the valley, leaving Delta to one side. When Root tactfully put it up to the Commissioners to make the decision they chose the eastern route.

The report also left a second point dangling—the method by which a channel could be maintained through the sand bars, constantly forming between High Falls and Carthage, sufficient to accommodate towing steamboats drawing up to four feet of water. Almost without exception engineers had favored the use of wing dams and jetties, in order that bar forming deposits of sand, as Jervis put it, might be carried by the stiff current through the narrowed channel thus created and lodged in the deeper water below.

The less favored scheme was for the construction of a dam in the town of Watson. This would set the water back to High Falls; below the dam the boats, it was held, would be in slack water to Carthage. The question thus posed Root also referred to the Commissioners. His report showed that the entire canal, from Rome to Ogdensburg, would cost $2,421,004.77 if jetties and wing dams were used; $2,431,699.29 with lock and dam. Neither estimate seemed to please the Commissioners. They were, as a matter of fact, so little worried that they dodged entirely the issue as between locks and jetties, and it was not until 1854, a year before the completion of the Black River Improvement, as the river part of the canal was officially known, that a decision was arrived at—and it was in favor of jetties! Work was at once started, but in 1857, three years later, the plan was again changed, and dams, not one, but two, each with its lock, were built, the first at Bush's Landing, in Watson, and the second at the mouth of Otter Creek. And that is how, standing on Beach's bridge, one can still see tips of a line of the old piles.

As for Root's final estimate of the costs, nobody had much to worry about for another eighteen years, when the completed canal, all the way from Rome to Carthage went into operation. And then the final construction figure was $3,581,954!

3

The Feeder Opened To Navigation

We are now in 1838, at the beginning of construction on the Black River Canal. It has been thirteen years since DeWitt Clinton first suggested such a canal, and seventeen more will pass before it is completed to Carthage. Thirty years in all. This fact, and the delays and the hesitations that caused it, are stressed in this account because they are of a pattern that extended on over into the administration of the Canal, from its opening until abandonment in 1922. Never in Albany did a clear-cut Black River Canal policy exist—never even a clear-cut theory of the functions of the Canal. Was the moving of goods its job? Or the carrying of Black River water into the Erie at Rome? Or a combination of both?

The State canal system as completed had four canals commonly designated as "laterals." They were, the Chenango and the Crooked Lake Canals, the latter completed in 1833 and the former in 1836; the Genesee Valley Canal, completed in 1857, and the Black River. All except the Black River Canal were abandoned in 1878, under legislative action taken in 1877. Reasons given for retention of the Black River Canal were blurred and consequently were impossible of serving as the basis for an aggressive, coherent policy of administration. The result was, to the end, utter confusion in the Canal's management. Although without Black River water the Erie Canal, which had been enlarged, could not operate, yet the vast sums of money spent on building and maintaining the Erie system of reservoirs at the headwaters of the Black and Moose Rivers were charged against the Black River Canal. The deficit thus showing on the balance sheet was used constantly to justify inadequate appropriations for maintenance, and as an argument by many, all the way down to 1922, for the Canal's abandonment.

These facts should be borne in mind as a kind of background for much that follows in these chapters. Meantime, there was action all along the line of construction. In addition to contracts let in Boonville in the autumn of 1837 for work between Rome and the mouth of the Lansing Kill, at Boonville on May 28th of the next year a public letting was held covering construction between the mouth of the Lansing Kill and Boonville, together with the Forestport feeder. The following September 5th contracts were let for most of the work between Boonville and High Falls, together with a few contracts, previously let, that had been abandoned or else not consummated at the previous letting.

By the end of 1840 the entire canal, from Rome to High Falls, with the exception of isolated sections, was under contract, the work progressing with rather astonishing speed. The contractors were happy, for they found that instead of having to haul in stone for the locks they could take it at nearby sites. In more than one instance lock walls were built of stone quarried from excavations for the lock pits. The hills back from the Lansing Kill yielded their full share, and, north of Boonville, the Sugar River quarries. The use of local stone was not always successful. Rock found along the Lansing Kill was often soft and porous, much given to chipping. Heavily loaded boats, and the seepage of water, were not kind to lock walls built of such materials, and when in the 1890's a good many locks had to be rebuilt, stone was brought in from the Sugar River quarries, until concrete later on was adopted.

The letting of a contract for any section of the canal was followed quickly by an influx of workers. Shanties for them sprang up where before were the sweet glades along the Mohawk and Lansing Kill; where a village already existed the newcomers merely attached themselves as temporary appendages. Such aggregations, while they inspired the starting of newer drinking spots, and added some fancy touches to the art of Saturday night brawling, at the same time played hob with the census figures and the dispositions of the men who gathered them. Being most of them Irish, and newly come over, the census had no place to put the canal workers except where it overtook them, whether it be

THE FEEDER OPENED TO NAVIGATION 31

in Western, Leyden, or elsewhere as each contract was working. Thus it was that the population of the town of Boonville in 1840 showed a step-up in population over 1830—from 2,746 to 5,519. In 1850 the figure was, to the embarrassment of the village boosters, only 3,309. It should be said, however, that the town of Ava was carved out of Boonville in 1846 and accounted for no small part of the population drop.

Regretfully it must be recorded that Irish canal workers stationed in Boonville as early as 1838 became involved, if by no more than a rumor that got around, in a down-river murder case. That the business ended in the first instance of capital punishment in Lewis County does not matter here; the important point is the high state of apprehension into which a community could be thrown by the name of McCarthy.

It seems that one Lawrence ("Larry" to one and all) McCarthy, living in the town of Watson, in Lewis County, had been charged with rubbing out his father-in-law. Trial in June, 1839, resulted in conviction, and August 1st was set for the hanging. Presently a rumor was spreading that Irish canal workers in Boonville were organizing an attempt to rescue their fellow-countryman from the jail in Martinsburg, that village being at the time the county seat. The rumor seems to have had some basis in fact; anyhow, to frustrate any attempts at rescue a volunteer armed group, dubbed the "Larry Guards," was set up, not only to surround the jail but to escort the prisoner to the gallows, the latter an assignment that they carried out with no little ardor.

What he feels to be an echo of the episode was heard by the author while collecting canal data in various parts of Watson. An older resident was asked about local fighters—men who might be famed in local legend for their prowess in the gentle art of socking, biting ears off their opponents, and similar manly pastimes. The gentleman could think of none in Watson, although he seemed to remember in the old days a pair of battlers coming down off the High Market hills now and again, most often to invade a dance hall, only to be routed ingloriously by some of the Merry boys. Suddenly he recalled hearing somehow his father tell of a band in his youth known as the "Lorrikins," a kind of vigilante band. Inquiries round and about brought substantially the same answer: Watson was always singularly free from brawls. In only one other instance in our studies had the Lorrikins ever been heard of, and then vaguely as the "Larrigans." Whether the term is a faint memory of the "Larry Guards" only more profound researches will reveal.

It is pleasant to record that all along the line of the canal many Irish workers, Operation Ditch ended, settled down in a community very much to their liking and became substantial, and often honored, citizens of the North Country. A letter of reminiscence written thirty and more years ago by a citizen, who as a young man saw the canal come to and go on through Boonville, was high in its praise of the workers who remained.

"The most numerous colony was, I think, the one about Hawkinsville," the writer declared. "A desire for a plot of ground that they could cultivate and a home of their own seemed to be the instinctive desire of each, and I think a wish to be their own landlord was what actuated them. But whatever the cause, none of them, except some who remained in the village, but owned and cultivated their own land and became self-supporting, prosperous and independent farmers."

Like the Irish the world over, the writer went on, "there were few teetotalers among them. None of them but proved their love and loyalty for their adopted country by becoming citizens at the earliest possible moment. Hospitable, courteous and generous, and the soul of humor and wit, they were ready in a moment to resent any insult or avenge an injury. I can recall, of those who settled in Hawkinsville, then called 'Slab City,' Dennis Monahan, Patrick and Dennis O'Brien, Patrick Hennesy, Timothy O'Leary, Patrick Sheehan, Patrick Dennis, William Buckley, Phillip Kennedy, and Michael Mulchi.

"Those who found permanent homes in the Lansing Kill district were John Holmes, Owen Maloney, Patrick Riley, Patrick Holmes, and, in its immediate vicinity on Jackson Hill and nearby, Owen Gibson and John Hennessy. Thomas Cross, Michael and James O'Connor, John Fallon, Timothy Hoban, Barney McLaughlin and Peter Campbell seemed to prefer the village life, although Campbell afterwards bought and moved onto a farm on Elm Flats. John Hyland, the shoemaker, Terrence Conlon the merchant, and John Black the tailor, all came about this time. John Leary, another of the sturdy sons of the Emerald Isle, who came with those spoken of, strayed from the others and settled on Potato Hill."

By the end of 1839 the State Engineer and Surveyor in a highly

THE FEEDER OPENED TO NAVIGATION

optimistic mood reported that, so well on was the work, $761,604.96 had been spent, $465,777.64 of it that year. This included work on the Black River Improvement and the Forestport feeder. At this rate, he was sure, the entire canal could be completed by October 1, 1842, that date being the beginning of a new fiscal year.

The flaw in the engineer's forecast was that in 1842 the legislators passed the famous "Stop Law." This act applied to all State canals under construction; the motive behind it was, the Legislature felt, the necessity of catching up with the State debts and maintaining "the State's financial integrity," as one of the proponents of the act explained it.

This move raised a storm of protest in that portion of the public opposed to the Democratic party. Silas Wright, of Canton, was coming up for the Democratic nomination for governor, and the Whigs seized upon the situation as an opportunity to discredit him and his party. Mass meetings voiced angry protest, editorials blasted the act, and manifestoes were issued, one of the latter ending its protest with, "Shall the system of public improvements go on, or be forever strangled?"

The act, however, stood, and a year passed before thought was given to the condition of the work that had been completed, or was well on in construction, before the law went into effect. Investigation by engineers showed that locks were beginning to deteriorate. The foundations of the locks rested upon a bed of timbers, and it was felt that exposure to the elements would bring rapid decay, and it was urged that means be found to give the foundations a protective covering of water. More than that, much timber contracted for had already been delivered and would quickly be worthless.

The Legislature accordingly authorized the sale of all delivered perishable materials, which meant principally timber. Nobody, however, wanted lumber, people along the canal route being, not buyers, but sellers of timber. As a matter of fact, $345.37 is the total amount received from all the sales made. Finally in 1846 the legislature authorized an appropriation of $2,500 to be spent upon preservation of the works—and only $1,000 of this was used.

But 1847 came, and with it the end of the Stop Law. The first act of the Legislature was to resume partial resumption of work, with special attention to rapid completion of the Forestport feeder and to putting locks on the Boonville summit level in shape to receive and pass on toward Rome, by way of the Lansing Kill, water taken from the Black River at Forestport. Thus urgent was the need of the Erie for more water.

Construction of the feeder caused engineers more headaches perhaps than any other part of the canal. Built along the face of steep sand hills, from fifty to seventy-five feet about the river level, it was subject, with sudden rains, to breaks in the tow-path, which was the river-side bank. As a matter of fact, during the Stop Law years several breaks had occurred, at the cost of much time and money in their repair. Fortunately, preceding 1842 construction had been rapid. The dam to replace the old Forestport dam, built originally to make a sawmill pond, was well under way, with most of the timbers for it delivered. Work on the guard-lock at the head of the feeder had progressed satisfactorily, and most of the timbers for it had been delivered; the guard-lock pit had been excavated, and the following spring would see the masonry laid—it was, and before the year had ended the feeder was finished and carrying water to Boonville for delivery to the Erie at Rome.

The first craft to navigate the feeder carried from Boonville to Forestport a party of inspection headed by Nelson J. Beach, State Canal Commissioner in charge of the Black River Canal. A description of the trip was published in the *Utica Press* nearly a half century later, giving, not only a pleasing picture of the trip but of the Forestport of the time.

"The authority for the statements herein contained," said the writer of the letter, a Forestportian, "enjoyed the first trip made by boat drawn by horses from Boonville to Forestport on the Black River Canal. The trip was planned and carried out by Canal Commissioner Beach on December 16, 1848. The craft was a small canal boat built at Boonville. Commissioner Beach invited a number of Boonville people to accompany him, which they did. John Utley, now of John Utley & Son, harness makers of this place, was one of the party. He states that the trip and weather were fine. The only ice in the canal was a little at Boonville in the wide water.

"When they reached Punkeyville they locked into the State pond, landing at the dock, and then, forming in line, marched to the boarding house near where King's blacksmith shop now stands, where they called for dinner, which was quite a surprise to the mistress of the house, Mrs. Sibyl Hovey, wife of George Hovey, who kept the boarding house at that time. They also visited the district school kept by Silas Hall, of North Remsen. Mr. Hall, thinking it quite a treat for the children, closed the school, and with the scholars went over to the lock, a short distance from the school house, to see the first boat and the privilege of taking a ride on the same. They were taken across the pond and landed at the dock in safety."

"Punkeyville" even then was giving way to "Williamstown" as a more dignified designation of the little village. Presently, however, the distinction of being on a river and at the same time the terminus of a canal feeder seemed to demand still more dignity and the term "Forestport" was fixed up. And "Forestport" it is on the 1858 Gillette map, although "Williamstown" occasionally appears in the public prints as late as the late 'eighties. "Punkeyville," however, is in current usage to this day by scoffers in their more joyous moments.

The place consisted, continued the letter, "of two or three houses, one of which stood where the M. E. parsonage is now located. A store and one or two houses were on the opposite side of the river, also two sawmills, one on this side, near where W. R. Stamburg's grist mill now stands, and connected with it was a tub factory. About this time, or soon thereafter, a steam sawmill was built on the site where is now located the plant of the Forestport Lumber Company. A large store was built on the corner of Division and Woodhull Streets, now occupied by George R. Ainsworth. The store was built at the same time as the first steam mill by Jackson & Blake. Mrs. Hovey is still a resident of the village. She is eighty-six years of age and remembers very distinctly the time."

Thus was the feeder, the first section of the canal, very prettily thrown open to navigation, and the first trip by boat described with a delightful nostalgic touch. Whether there would ever be more was in the lap of the gods, or of the politicians in Albany. Fortunately for the people of the Black River valley the gods were on their side.

4

Boating Starts

Actual building of the canal was not a snail-pace business at that. True, from 1838, when work got under way, until May 10, 1850, when the first boat, swinging out of the Erie at Rome, reached Boonville with a full cargo of merchandise for the merchants of the region, twelve years had passed. From these, however, five years of work suspension under the "Stop Law" must be taken, leaving seven years for actual construction. In these years twenty-five miles of canal had been completed, and most of the work between Boonville and Port Leyden.

Building the canal was vastly more than just digging a ditch. Especially up through the Lansing Kill Gorge, where, in order to maintain levels between locks, the canal had often to be built on top of fills. As if this were not enough, thousands of yards of dirt had to be hauled in to build up the towpath, which also was the canal's outer wall—for here, as with the Forestport feeder, the tow-path was on the stream side of the canal. A hundred feet above the Kill bottoms the canal clung perilously to the sides of the hills, in places the drop almost sheer.

Such was the stretch of canal which ran from the Five Combines practically to Lock 64, where the State road crosses the canal three miles below Boonville. But building the channel was only part of the job. There were the locks — seventy-one of them between Boonville and Rome. The amount of labor involved in building a lock can best be seen in the fact that when in the 1890's the State rebuilt some of the old ones, a single lock meant an all-winter task, costing as much as $25,000.

Then, too, each lock required a sluice for carrying surplus water down around the lock into the level below. For these, run along the heel ("berme," should readers want to be classical about it) bank, a channel had to be made, and in some cases, where the drop was precipitous, the banks of the channel needed walling. With what care a sluice must be constructed may today be observed at the remains of the Five-Combine locks, where the water rushed down a decline of fifty feet.

There were aqueducts to be built, too, stone structures to carry the canal across streams, four of them. One ran over the Mohawk just above the site of the Delta dam of today; another was at Frenchville for the crossing of the stream at various times known as Wells and Willis Creek and Big Brook. The third was at North Western, over Stringer's Brook, and finally the aqueduct just below Dunn Brook, which carried the canal over the Lansing Kill. The ruins of all, save the first Mohawk crossing, where the Delta reservoir covers them, may be seen hard by the highway, in their every remaining detail eloquent testimony to the painstaking work which went into their construction.

And equalling the aqueducts in sturdiness were the four culverts. These were a major operation, for under the canal and into the Lansing Kill they carried creeks that, in summer peaceful and well mannered enough, yet by freshets were turned into rampaging torrents.

And there were the seemingly interminable bridges. Not only bridges carrying highways over the canal, but bridge-crossings on farm roads as well. The engineers' reports list thirty-nine of them, four of them in Rome—at Whitesboro, Dominick, Stanwix and Thomas Streets.

Construction of the Boonville bridge at Main Street, with its wide sweep over the canal, high up to permit of the passage of boats underneath, held a fascinated public under a complete spell. All evidence points to the fact that the famous Committee of (to put it mildly) Forty-Seven worked overtime in giving unrelaxing attention to every detail of the work. And evenings, when the Committee knocked off, the younger set took over, romping over the stringers and daring and double-daring the more timorous of the small fry.

One prank nearly ended in disaster. An all but eye-witness account of the lark, written years afterward, deserves quoting.

"The bridge," said the writer, in a nostalgic vein, "was constructed on the plan of all old-fashioned bridges—a double track railing, with a railing between of the same height as the outside barriers. A party of young people was standing on the bridge one night, in a beautiful moonlight, and the question was raised as to the possibility of jumping from the top of the outside railing to the middle one. Of course femininity screamed with one accord, 'It's impossible, it can't be done!'"

Presently a young lawyer, Ball the name, found himself somehow the object of the rising clamor. In a "What have I got to lose?" spirit he accepted the challenge implied in the young ladies' cries. Mr. Ball "declared it could be done easily, and immediately prepared to accomplish the daring feat. He did, and more: he went directly over the top of the division railing, not having observed the precaution to look before he leaped. As there was no flooring on the other side of the division he went down and down, landing upon the rocks in the channel below. His companions, terror stricken, rushed to his rescue. He had fainted, and was carried, soiled and bruised, to the hotel, where he was confined many weeks. He was the hero of the hour, and cheered by the attention and bouquets which he received."

The story ends on the sad note that as soon as the young man had recovered "he closed his briefless career in Boonville and sought more genial climes." To be more specific, he went to Texas, where "his remains repose down by the Rio Grande, but the bridge was called 'Ball's Bridge' for many a long year."

Bridges had been around for a good many years. They had served the age-old purpose of bearing traffic over rivers, and affording a vantage point to young swains who on them could the more readily point out to the fair one at his side how like to her eyes were the stars shimmering on the water below. Bridges over a canal apparently were something else again, the way the young folks were always taking liberties with them. It remained for a pair of young Hawkinsville bloods to think up the year's most original idea.

The tale is of how these two learned of a call to be paid that evening by the village's most romantic youth to his best girl over back of Hayes' Corners. Come dusk they climbed the hill and removed some of the planking from the canal bridge. Then they removed themselves from the scene. The young lover, rigged out fit to kill, a half hour later mounted the hill, unaware, as the story in years after used to be told, of the fate that awaited him. He took a half dozen steps onto the bridge, but the next one or so he never completed; like Mr. Ball, he went down and down, on the way catching himself a stout clip on the chin as he passed a plank, wrenching a ligament loose as he clawed at a stringer, and emerging, according to the story, a sadder and a wetter man.

Hawkinsville people, it may be added, had a peculiar right to be fresh with bridges. Did not one of their neighbors, Stephen Potter, build all of the original Forestport feeder bridges? Starting out as a farmer he had advanced, if that is the word, into the contracting business. Among other structures of his which are recorded was the old chair factory at Hawkinsville and the horse tramway built to carry logs from Grantville to pond-side for the Forestport mills. Potter it seems took bridges in his stride for in 1901 he died in Boonville at the ripe old age of eighty-nine years.

While the Boonville bridge was, with the dubious aid of the Committee of Forty-Seven, being put together, other excitements up and down the village-side of the canal were going on in the place. The air was filled with the sound of saw and hammer; new businesses were coming into being, needing new places to house them.

First to break ground was the partnership entered into by Jesse Talcott and Whitmore Hall, which set itself up as a forwarding house. The situation seemed propitious for such a service, since the canal was bound to be followed by a lively buying of butter, cheese and such other produce as the region would be producing in increasing quantities, now that they could be got to market. Commission firms would spring up and need warehouses and shipping facilities, and that is where Jess and Whit came in.

Unfortunately, while several of the retail merchants of the village did patronize the new concern, yet many of them one by one built their own warehouses, or "stores," and the historian is not surprised when he learns that after four or five desultory years the outfit was taken over by Peter B. Post.

The Holdridge & Gilbert foundry followed shortly, its building rising immediately north of where Main Street slipped over the Ball bridge. Then followed the enterprising firm of Joslin & Shultz, chief of the local retail concerns to cut into the territory which Talcott & Hall had staked out for itself, putting up a warehouse just north of the foundry. It may be noted here that in course of time Sam Joslin dropped out of the business and Billy Davis came in with Bush Shultz. The only reason for mentioning the matter here is that the narrative might record the fame brought to the concern by Phillips' muscular feats. The most spectacular of these was to take a barrel of salt from the ground and lift it high up over the sideboards into the wagon-box of a waiting customer. So spectacular a display of might would alone explain the fact that this forwarding house, specializing in butter and cheese, was one of the most prosperous ones in the village, and long continued to flourish.

Then just beyond Joslin & Schultz was the store being put up by Orsemus Clark, who among other claims to a solid fame was a successful hitch as proprietor of the celebrated "Hemlock Store" up on Potato Hill. Meantime over across the canal Gilbert Brinckerhoff was fixing to install a grocery stock in his home for the convenience of boatmen.

All these doings were preparatory to the opening of the canal, which so far as Boonville was concerned, was going to get off to a flying start.

Of the communities between Rome and Boonville North Western most sharply felt the impact of the canal. Prior to the beginning of construction the term even of hamlet would have exaggerated the importance and size of the place. With the coming of the canal laborers, however, one Bissell, up from Rome, started a small store, which he stocked with goods suited to their needs, not omitting vast quantities of Good Old Summertime, Jolly Tar, Peiperheidsick and other brands of chawin' and smokin' favored of canallers. This, scarcely more than a shanty affair, was followed by a more pretentious store kept by David Brill, a local cheese maker of importance, who in the year of the canal opening built the Half-Way House, to this day known by all travellers up and down the valley.

The man who, for sheer energy and versatility, made the village (for by now Mr. Brill, who owned much of the immediate region, had platted the place) distinctly a canal town was Jerome V. Gue. Gue as a lad of fifteen had come here from Boonville just in time to meet the canal coming up. He got a job as carpenter. In good time he went into business for himself, and on the southern edge of the village may still be identified the site of the drydock which he was to build and own, and near-by the location of his boat-building yard, cider mill and other of his enterprises. His boats, while he did not build many, were regarded by boatmen as among the best in the service.

The completion of the canal brought a need of men, not only for new businesses like the Gue activities, but for maintenance and operating jobs like carpentering and lock-tending. Consequently a steady migration of families from the hills of Steuben and Western got under way, to the extent that few are the families of the village today that are not tied by tradition and background to one or another of these towns. North Western was decidedly on the canal map.

In a minor way the same processes were at work all along the route of the canal. Down from the hills came the Dorns and the Davises and the Clarks and the Van Dewalkers, to note but a few of the names that to the end were to be identified as canal families. And as the number grew they shook down into a number of small settlements like Hillside, Dunn Brook, and Hurlbutville, with even smaller clusterings that never so much as achieved names.

All this the canal did, and, though at a slower tempo, it is what it did to Forestport. Forestport may be said to have started from scratch on its journey to eminence. More than that, it was handicapped by the name of Punkeyville. And yet by 1858 it could boast of a blacksmith and machinist's shop, a State canal house, a school, a hub-and-spoke plant, two hotels, and three stores—one a general and another a grocery store, and the third a combination grocery and saloon. All told that meant three places where a citizen could slake his thirst.

Yet to come were the men who were to make of the village and its out-lying forests one of the State's most exciting lumbering centers. The roll of them was to be a long one, and in it would be the McGuires, the Stamburgs, the Dentons, the Waterburys, the Hoveys, the Harrigs, the Sypherts, the Gallaghers, and many another. As of 1850 that was several years away, but when it came it was the canal that had made it possible.

Such was the state of affairs to which the skipper brought the little Joslin & Shultz boat that afternoon of May 10, 1850. It was not much of a craft, one suspects. Little is known of her, since Black River boats, for some reason never explained, were not registered. The known facts are that the partners had found her somewhere down along the Erie, probably at Rome. A deal was made, and, now bearing on her stern the brave name of the firm, she was sent to New York for a cargo of goods which in good time found shelves awaiting them in various of the Boonville stores. A few days later she pulled out of the Boonville basin for New York, bearing a cargo made up mostly of butter and cheeses, stowed away under the careful scrutiny of Whitmore Buck. It was Whit's first experience as boss stevedore, but from all reports it is clear that never was a loading job done better.

5

Whistle 'Round the Bend

Boonville had no sooner recovered from the excitements of greeting the canal than, one morning in early November (that would be 1850) word spread throughout the village that the gates of Lock 71 were being opened. The level to the north was going to be filled. The Committee of Forty-Seven dashed to the scene, only to find that a dozen canal engineers had got there first, at their head no less a person than that efficient Resident Engineer, Daniel C. Jenne. Jenne's was the job of completing the canal on north, just as five years later he was to bring to Boonville the Black River & Utica Railroad, to become known up and down the line as "The Sapbush."

The Canal that November morning had just been completed to Port Leyden, and while boats would not begin to run until the following spring, yet water must be eased into and through the various levels and locks in order to ascertain whether they would hold. At Port Leyden the water would be turned back into the Black River until such time as the canal would be continued to Lyons Falls.

Several days would be required for this testing operation. The full volume of water would not be turned on until the dirt banks were thoroughly saturated and possible weak spots discovered. Too sudden and too great a volume would be too much for the dry canal walls and possibly would result in "breaks."

At last the probing and the testing were ended, and the result, history is pleased to record, was gratifying to all concerned; next spring the canal was opened to navigation as far as Port Leyden.

Port Leyden was a pleasant place to halt. As yet it did not have much to offer in justification of its new name; it was now, however, a canal port, with a small hotel or two, a sawmill, and at its very door was no end of natural resources that in a few short years from now were to convert the village into a hive of industry. All that would start to come with the discovery of iron in its midst in 1864, the rise of its tanneries to the stature of big business, and the cutting over of the timber lands just across the river. Meantime it concentrated on the produce of the adjacent farming regions, and fed a steady stream of butter and cheese into the canal traffic.

Not that Port Leyden couldn't go on having dreams of eminence, or have some one else do some dreaming in its behalf. As when agitation was started in Boonville for a national bank. Promptly the Lowville *Journal* put Port Leyden forward as a likelier place for a bank. "From its situation on the canal and river, with its water privileges," said the editor, "it is destined soon to become a large manufacturing town rivalling that of Boonville. With all these prospects in view would it not be for the interest of the business men of Port Leyden and vicinity to use their capital in establishing a national bank at that place instead of Boonville? Is it not a matter worthy of consideration that their capital be kept at home, in their own county? Think of it, gentlemen!" The *Black River Herald* in Boonville reprinted the suggestion without comment, regarding it probably as a kind of Wimpy "Let's you and him fight" business. Anyhow the business ended there.

Port Leyden ceased to be a canal terminus in 1856, when navigation was opened to Lyons Falls. Here again traffic must virtually end for the time being. Lock 109 could pass boats down into the river, but just now there were no adequate towing craft to take them through to Carthage and intermediate points. This situation was changed the following year, when Lyman R. Lyon built the "L. R. Lyon," a sternwheeler drawing fifteen inches of water. Without channels, however, a boat with even so light a draft found the trip hazardous during the dry summer months. Work of canalizing the river was a long way from complete, and after many changes in plans people were wondering if the work ever would be finished.

Now some of the canal engineers from the very first had been in favor of canalizing the river by means of a dam to be built, with lock, at Otter Creek. This they said would set the water back, to a navigable depth, as far as Lyons Falls, while a dam to be built at Carthage would give slack water to Otter Creek. A majority of the engineers, however, favored the plan of building so-called "wing dams" at the head of stretches where bars were in the habit of forming. These dams, also called jetty dams by the engineers, would force the water to flow between parallel rows of piling. The new channel created in this way would be five feet deep and wide enough to accommodate towing steamers. The increased flow of water created by the narrowed channel would discourage bars from forming, but, as an auxiliary aid, drudges (which was good canal language for dredges) would be kept at work.

The latter plan had been adopted in 1851, and contractors were soon at work accumulating men, materials and machinery, Big Three in the language of the canal construction men. The next year, however, all work was called off when the Court of Appeals decided that the law under which the contracts were let was unconstitutional.

The comedy entered upon its second act in 1853, when the Canal Board went back to dams and locks, only now there would be two of them, one at Otter Creek and the other at a point east of Lowville to be selected by the engineers. Contracts were promptly let, only to be cancelled the following year, this time because of fraud discovered in the lettings. December came (that would still be 1854), and—again the plan was changed back to jetty dams and piles.

This one was taken seriously, so seriously that the State Engineer and Surveyor could report, as of the end of 1855, that on the 16th of the previous February "about eight miles of the improvement of Black River, below Lyons Falls, were under contract, including the snagging of the river from Lyons Falls to Carthage, and on the 24th of August the balance of said improvement was put under contract." The contractors, he went on to say, "have expended a large amount of money in making preparations for the prosecution of the work. One steam dredge was put on in the early part of the season, also a steam pile driver and other necessary boats and fixtures for carrying on the work." Later on two more dredges were added, and big things were contemplated.

And so it went until 1857, two years after the coming of the canal to Lyons Falls, when the Canal Board again reversed itself and went back to the dam and lock idea. Up to this point $113,751.49 had been spent on the work. The expenditure of $9,134 more, on the basis of estimates, would, according to a statement made by Jenne, have completed the whole enterprise. Otter Creek was again chosen for the upper, and Bush's Landing for the lower dam. The former was completed in 1860, but the latter only in 1866.

And that is how it was that the river part of the Black River Canal in 1856 was far from entirely ready for business. But boats could, and did, get through. One skipper was heard to complain that it was the army of engineers poking around, and the drudges and workers on the dams, more than the sand bars and dead-head logs, that were the boatman's greatest hazard. The fact is that before the "L. R. Lyon" was launched in June, 1857, some boating was getting through; in May five "canal barges" were reported as leaving Carthage, southward bound, though what and who towed them is not known. With the completion of the Otter Creek dam and lock in 1860, however, canal traffic for all practical purposes had finally arrived.

In the records of the day there are no reports of undue rejoicing along the river upon the arrival of navigation. The citizens had got along without big-time traffic for these sixty and more years, and they took its coming without transports of undue emotion. There it was, however, and to it one and all proceeded to relate themselves with commendable zeal. One of their earlier interests came with the discovery that traffic on the river and highways was increasing each day, and that before long hotel accommodations would have to be provided. With this discovery there came into being a number of hostelries, the most celebrated of which was opened by Jack Pasenger near the east end of the dam at Bush's Landing, to be best known as "Jack's," or "Pasengers," and in course of time "Pasenger's" in the public prints of the day came to be used to denote even the village itself.

Pasenger's can be said in all truth to have been an institution. Its proprietor was immensely popular. He was a huge man (there were nearly seven feet of him), and he had a booming voice that, when occasion demanded, reached far across the river and over into the town of Martinsburg. Legend has it that it could be heard, on a fair day, as far up-river as Pine Grove.

Legendary or not, it was a very real voice that astounded a delegation of canal brass who, later on, on a tour of inspection from Rome, were having a pleasant voyage along the river on one of the finer craft, the "J. F. McCoy." In the party was the State Superintendent of Public Works, H. B. Dutcher; the State Engineer and Surveyor, Horatio Seymour, Jr.; the Superintendent of the Black River Canal, Sam Ferguson, of Boonville, and prominent local officials and citizens representing practically a Who's Who of Lewis County.

From time to time the officials underwent the ordeal of climbing off the ship to look over the State's dams and locks and bridges, which was the cue to the home contingent to make for the bar rigged up for the day. Unfortunately the schedule did not include a stop at Jack's for refreshments; instead aboard ship a sumptuous luncheon had been planned, prepared and was served under the keen eye of George May, of Boonville, the hustling hotel man who had bought the Hulbert House in 1870 and was always doubling in something, just now as clerk to the collector of tolls in Boonville.

Due stop was made for the inspection of the Bush's Landing dam and lock, situated some twenty rods or so up the river from the hotel. At the precise moment that the "McCoy" emerged from the lock and was getting under way the mountainous figure of Pasenger loomed up on the shore and from across the waters the voice boomed, "Hooray for Hancock!" Jack suspected that every canal official was by definition a Democrat, and though the Fremont presidential campaign by now was history Jack was having his evenin's. Charley Chickering, of Copenhagen, the district's popular Congressman, was in the party and in an after-dinner speech at the Levis House in Carthage that evening he declared that Jack's dramatic performance alone was worth the trip.

Pasenger's was a convenient, and a favorite, rendezvous for boatmen, for men going into and coming out of the Stillwater country, and for the lumbermen of the region. Every inquiry of old boatmen as to possible carryings-on in the hotel brings the same reply—Jack stood for no brawling. His kindliness and good nature were outstanding traits. Mrs. Rinkenberg, of Port Leyden, remembers as a child riding along the road through Bush's Landing with her father, who was John Hodge. Always the horse would stop in front of the hotel and refuse to go on until Jack came out with a lump of sugar or a piece of candy.

One of the stories Jack used to tell with enthusiasm has more meaning for rod-and-fly men today than it had even in Jack's time. The story had it that a couple of State canal engineers, sent into the forests beyond Number Four to find a likely reservoir site, came out with a catch between them of some six or eight dozen trout, and declared that they would never, so help them, go into that God-forsaken country again.

Scarcely less celebrated than Pasenger's was Wellington Brown's hotel, which stood on the east bank of the river a mile above Tiffany's Bridge, the site of the Glenfield of today. To this day the place is remembered with affection by the older of the old boatmen. Pasenger's was burned in after years, but the center of the Brown establishment to this day remains, a farm house, the first on the left after leaving the village. The photograph makes clear how pretentious a place it must have been in its time. The Saturday-night dances, held in the wing to the left, were extremely popular events—less popular with the horses which carried the guests, one surmises, being quartered for the evening in the shed below the dancers and longing probably for a more quiet zone.

Wellington was the only inn-keeper in recorded Greig history who had four wives. Somewhere along the line one of them became irksome to one of her many step-daughters, one of whom gave an eager yes when the youth of her choice asked would she marry him. The assent of her parents she was sure would never be given. Especially since the suitor lived in Carthage. Anyhow, not taking the trouble of asking they instead planned an elopement by canal. Which would make it, no doubt, the first, and perhaps the only, canal elopement in American history. In other parts of the world, notably in Venice, a canal is of course a necessity if one is going to elope.

In their carefully laid plans the young lady on the day set was to listen for the whistle of the approaching Carthage-bound steamer. This being a whistle spot, the young man went in advance to the captain and got his promise to whistle and stop, whether or not there were passengers to get on or to get off. Meditating later upon the faltering nature of human memory he himself on the appointed day was in Lyons Falls and boarded the down-boat. He stood, as they approached the bend in the river, at the captain's side to make sure of the whistle.

The whistle did sound, and the young woman, looking her dainty prettiest, slipped out of the house and scampered down the steps and across to the boat. The captain was so in accord with the spirit of the hour that no time at all elapsed between her first footstep on the deck and the movement of the ship downstream.

The secrecy with which all of the plans were made strikes the listener to the tale as being beautifully contrived. And yet, when at the end he learns that every tree in the vicinity was filled with kids, in from the neighborhood to see the show, one wonders!

Not content with running a hotel, Wellington got himself listed in the armed services in the War between the States, and his son Charles, sixteen years of age, with him. He served in the 186th Regiment, New York Volunteers, Company F, in which he held the rank of orderly sergeant.

The arrival of the canal inspired no new hotels in the Boonville-Lyons Falls sector. The fact of the matter is that Lyons Falls had jumped the gun as long before as 1850, six years before the opening of navigation there, when Gordias Gould built a place which he named the McAlpine House, in honor of William J. McAlpine, who next year, and for two years after that, would be the State Engineer and Surveyor, and by virtue of that office member of the State Canal Board. In a canal-side setting of rare beauty, high above the river, the McAlpine earned for itself an enviable reputation for the excellent quality of its appointments and service.

To have made necessary this, and the places put up by Jack Pasenger and Wellington Brown, was good going for the little canal, as yet not even finished. This would be only the beginning.

6

The Breaks

Wives of boatmen who became too expert at such strenuous pastimes as checkers, seven-up and the like blamed it mostly on canal "breaks." When canal walls gave way and the precious water rushed through the opening out over the surrounding terrain, the breach might be a boat-length, or as much as three hundred feet long, and boats grounded for twenty-four hours to two and three weeks. And if it wasn't breaks it could be sink holes. On these latter occasions, less frequent than breaks, the bottom of the canal simply disappeared from view.

If the break was in the southern stretches of the Lansing Kill, or along the Mohawk, the crew from the State shop at North Western would swing into action; if in the upper regions the Boonville crew. For the canal, for administrative purposes, was at first divided into two sections, each with a superintendent and a State shop, manned by a force of carpenters and laborers, and with a crew for a State scow. At a later time the feeder had its own maintenance crew, stationed at Forestport.

Section 1 ran from Rome to Lock 70. Section 2 started at 70 and extended to Lyons Falls and included the feeder. Later on, when navigation was continued on to Carthage, the river, from Lyons Falls on north, constituted Section 3. The arrangement turned out to be pretty fluid, for at various times the Boonville level, up to Lock 71, together with the feeder, were put in Section 1, while at other times Sections 2 and 3 were combined. Changes were so frequent, as a matter of fact, that Ed Ford, of Forestport, used to say he couldn't keep track of them. Down on the Erie one time with a load of lumber for Albany, a stranger asked Ed where he hailed from. "I don't know," he replied. Asked how come Ed grinned, "I fergot to find out when I got to Boonville." Telling the story when he got home, Ed said "the feller looked at me fer a minute then walked away fast-like."

It didn't matter much to a stranded boatman from what section the repair outfit hailed. He and his family (who in as many cases as not travelled with him) were used to these interruptions in traffic and just settled down for the duration.

If the break struck while a number of boats were waiting to be put through a lock a little colony would be thus formed. This would give the men-folks a chance to fix up the broken hamestrap on the harness cf the off horse, catch up on their whittling, and swap yarns and bits of news up and down the canal, such as how Art Lane had got himself a new boat, the "Doris E. Lane," and the strange happening that Roscoe Clark brought back from down Greig way—about how the carriage in Abbey's sawmill, Roscoe's last trip down there, at the end of the run through the saw had something go haywire with its under-pinning or something. Anyhow the carriage when it was put in reverse wouldn't stop and, gathering momentum as it went, it came to the end of the track and shot out far over the bank, landing in the pond, startled setter, log and all.

The tie-up would be nice for the womenfolk, who could visit back and forth, and talk about how long ago it was they had last seen one another, Mrs. Tom Barnes's telling how her son Willie got the measles down on the Erie and they had to get in a doctor from Sprakers, and Mrs. Andy Beach giving a history of their last trip to Troy, where she bought some new curtains for the cabin windows of the "Jennie Beach" and took little Edith, carried on her father's shoulders through the crowd, to see Barnum's, her first circus.

The young fry of the new community would be doing all right by themselves, too, especially if the break had occurred along the Lansing Kill, where almost any place they could get out their fishing gear, dig up some worms, slide down the creek bank and, hoping for trout, come up with a mess of horned dace and such. They could have got eels in some of the locks in the sixty numbers. Here lock tenders picked up a penny now and then by selling eels. The quickest way about this business was to throw a line in just back of a boat. Here the slimy ones congregated to pick up bits of food dropped overboard from the boat. The lock tender after landing an eel felt his day completely lost if he did not turn around and sell the creature to the boat which had acted as a decoy.

It would be fine and dandy, of course, if a couple of boats were neighbors back home. This gave the whole family time to check over recent doings at home. If it was back in '63 it might be George Parrish, of the "Major Anderson of Fort Sumter," and Ed Regan, of the "John and Susie," both of Forestport, hung up at 46. It would be the first time George had seen Ed for a month or so, and after they had taken General Grant's campaign at Vicksburg apart, George would ask if it was true what he'd seen in the paper about a visit Ed had paid to Boonville. Ed would be all innocence as George pulled a crumpled clipping from his pocket.

"Here it be." A grin all over his face, George read, having a hard time with the big words. "It says here, Ed Regan was weary of the slow an' easy motion of his canal boat and started to take a ride on terry firmy in one of Hank Cramer's best vehicles. He drove the nag and buggy up before Hulbert's hotel and made fast to a snubbin' post, then an' there standin'. On approachin' rather abruptly to cast off the bow line, the nag unshipped its head-stall an' proceeded to move off rather fast, whereupon the quadruped resented the familiar embrace and proceeded to gyrate in such a manner that the centrifugal force soon threw the boatman off in a tangent, whereupon the nag began to execute sundry evolutions and hasten towards the stable with a careless speed, which soon reduced the buggy to a wreck, to the great detriment of shade trees, sidewalks and snubbin' posts. No one was hurt."

Ed listened attentively to the end and after a minute he said he guessed as how there might be jest a mite of truth in her.

The problem of water in its various manifestations was never very far away from the boatman. If it wasn't the breaks that every year cut the navigating time short by weeks, there was the annual freshet season, during which creeks coming down off the hills of Ava, Western and Boonville clogged their culverts and poured into the canal levels hundreds of tons of shale and other debris. This had to be removed, and like as not holes torn in the canal banks repaired, operations that might run into weeks. Locks located immediately at the bottom of steep, high hills, like lock 33, and the Five Combines, lay wide open to quick rains. In March, 1865, practically the entire tow-path between 32 and 34 went out, and the reinforced wall put in by the engineers failed to give it immunity against too frequent breaks in the future. Even in the lower, comparatively level lands around Delta the canal suffered from freshets. On one occasion the Mohawk went on a rampage and broke through a dam which was utilized to turn Mohawk water into a feeder across from Delta to the canal below Lock 9. It was a privately built and owned affair; the owner refused to make repairs and the State was forced to go in and put up what amounted to a new dam.

Heavy rains sometimes loosened the steep shale hillsides in the Lansing Kill section of the canal, causing destructive slides that held up traffic until the debris which they shoved into the canal levels could be cleaned out. The account of some of the more disastrous of these slides, and of the more spectacular breaks, belong in another chapter, and here may be set down the story back of another set of causes which made for long and costly lay-up of boats. Referring to low water in the Black River in summers so exceedingly dry that there would be insufficient water for either the canal or the mills along the river.

The engineers from the beginning had calculated that 16,000 cubic feet of water every minute would have to be taken out of the Black River and fed into the canal at Forestport to insure uninterrupted navigation. Of this amount approximately 4,000 feet would be taken out of the Boonville level for northward traffic and put back into the river at Lyons Falls; 11,000 feet would be necessary (and it was held that this would be sufficient) for navigation between Rome and Boonville and for feeding into the Erie Canal.

The estimates, however, were found to be on the under side. The measurements had not been made in exceptionally dry midsummers to ascertain if the river at such times could supply 16,000 feet to the canal and leave any water at all, save the 4,000 feet returned to the river at Lyons Falls, for the mills below Forestport. As a matter of fact, as stated above, there were many summers so dry that for long periods mills were forced to shut down for lack of water. Out of these experiences people owning water rights along the river began to enter claims for damages against the State that reached formidable figures—many, who did not own mills, said too formidable. Claims came from as far away as Watertown, where important mills were completely paralyzed. Some manufacturers managed to keep going on reduced schedules by

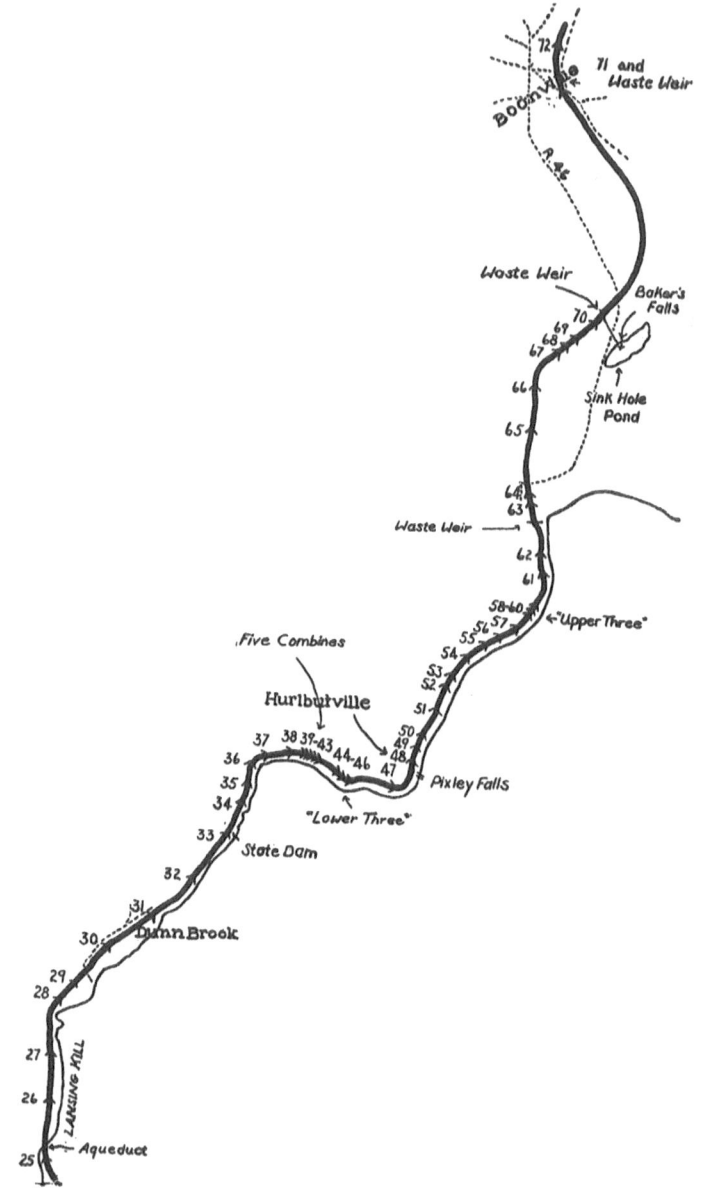

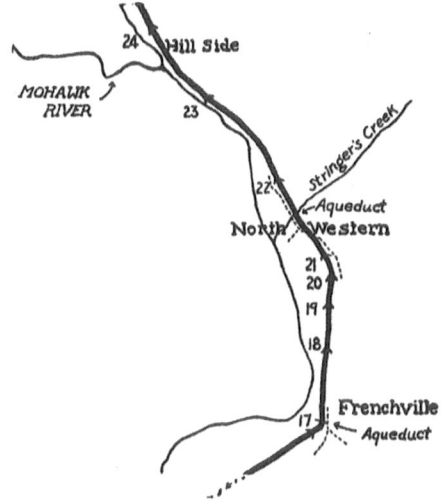

Map of the "Alpine" Section of the Black River Canal, with lock numbers shown. The Boonville-to-Rome highway is not shown, except from Boonville to lock 64 and through villages, inasmuch as it follows, most of the way, a line parallel with the canal.

rationing themselves and shutting the gates of other mills to which they sold water.

These people finally rose in their wrath and inundated the legislature with bills of grievances, and in 1855 the Senate appointed a committee to look into the situation. Starting along the Lansing Kill they took depositions from mill folks all the way down to Watertown.

The deposition made by Jesse Babcock, of Dexter, is typical of the rest and, being shortest, may be given here in substance. Jesse deposed that he had been a "resident of said village since the year 1837; that during all that time has had occasion to use the water of the Black River for hydraulic purposes, with slight exceptions; that until the summer of 1849 there was an ample supply of water for the business carried on at said village, every season, and at all periods of the year, especially after reconstructing the old dam [in Dexter] in 1846; that for the past nine years, or since the rebuilding of the dam, there has been an average of about thirty wheels . . . engaged in the woolen manufacturing business, sawing, grinding of grain and plaster, planing of boards, cloth dressing, sash making, wheel wright business, and cabinet making. That in the summer of 1849 an extraordinary deficiency of water was for the first time experienced, and with the exception of the present year, 1855, and to a reasonable extent in 1850, there has been a diminution of water power fully equal to one-half; that the woolen factory at Dexter holds the first right to water from the dam, with one slight exception; and that in 1849, as well as in 1852, '53 and '54, the said woolen company required all other gates to be closed, to such a degree that not a single grist could be ground without a special permit." And this meant, said the deposition, that if things continued thus more than one-half of the invested capital would be proved a total loss.

In the Lansing Kill valley, on the other hand, the committee found another story. Here the deponents were happy about the whole thing. They now for the first time were getting water, and to spare, for their wheels. As when Peter S. Baker, of the town of Boonville, deposed that he had resided there since the year 1817, "and has occupied the farm on which he now resides, two miles southerly of the center of Boonville village, for the last twenty-one years; that in the year 1849 the State let water into the Black River Canal, by which means a water power on his premises, which could previously be used only at times, from the adjoining swamp, was made ample throughout the season, when the canal was in operation, or the water suffered to flow therein during the winter months, as has been the case for two winters; that the water power added by the State has enhanced the value of land all the way from the first sluice near Lock 70 to Rome, wherever the make of the land would afford conveniences for mills to be placed."

The firm of S. C. Hurlbut & Son was located at what later came to be known as Pixley Falls, but then as Hurlbutville, a name by which it is shown on the maps of the period. W. W. Hurlbut's deposition had much the same story to tell as Baker's. His father and himself had, previous to 1852, resided in Herkimer County, where they had engaged in the lumber business, running a sawmill and making cheese boxes. He then goes on to depose that hearing of the improved water power on the Lansing Kill they "purchased one hundred and twenty-five acres of land on both sides of the Lansing Kill, at what was known as 'Lansing Kill Falls,' with the design of using the hydraulic power in part themselves, and of selling to others; that it has been represented to them that the water power on this place, previous to the introduction of water by the canal and feeder, was adequate only to the operation of a single saw gate or mill; that since their purchase and occupancy the increase of water power is from two to three hundred per cent; that the value of land at the time of their purchase, in the vicinity ranged from thirteen to twenty dollars per acre, and previous to the construction of the canal from four to five dollars per acre; that they value their real estate at Hurlbutville at the present time at ten thousand dollars."

The report made by the committee brought action in one important direction. In the early years of their troubles the users of water had entered claims against the State for damages, and the Canal Appraisers had allowed them. Suddenly all payments were stopped by the Canal Auditor, on the grounds of a strained, so it seemed to the North Country men, interpretation of the law under which in 1836 the canal had been authorized. Upon recommendation of the committee damage payments were resumed.

Other relief also was in sight; the State had begun developing a system of reservoirs for impounding water for summer use throughout the headwaters of the Black River. This was in accordance with plans suggested by engineers as far back as 1825. Surveys for possible sites for dams in the upper Moose and Beaver country were also being made. As a matter of fact, the North Lake reservoir was even then nearing completion, and was put into use two years later, in 1857. Its capacity was 337,851,360 cubic feet of water. By 1882 the Black River system had reservoirs also at South Lake and Woodhull, Sand, White, Canachegala, Twin, Chub and Bisby lakes, for a total storage capacity of 2,091,763,120 cubic feet.

Even two billion feet of water, as it turned out, was not enough water during dry summers to keep boats running without interruption.

7

Fun On The Canal

While the canal was built as a carrier of commercial and industrial products, yet the activities conducted upon its fair waters included programs that catered to the lighter inclinations of its public. First among these was the toting of picnic parties from one point to another. The earliest of these affairs of which we find record, although earlier ones undoubtedly had been held, occurred on September 15, 1858, and was participated in by the Sunday-schools of Christ Church of Forestport and Trinity of Boonville. On canal boats which cannot now be identified, the parties met at Miller's Mill, or Miller's Grove, halfway between Hawkinsville and Forestport. Next in interest to the eating part of the day's program is the story of how the Christ Church young people, as the Boonville party hove in sight, lined up in a double row, through which the alighting young people from Trinity passed on the way to the grove.

The latter spot, according to the *Black River Herald* account, "was certainly one of the very best. The audience was seated in an area surrounded with a thicket of small trees, which formed an agreeable, cool shade. The platform occupied by the clergymen was erected in a recess—in front of the audience—and was entirely overhung with boughs, the teachers, children and friends being seated, and preserving that decorum and good order which marked their conduct through the day."

After an interval of singing came the feature attraction of the day—which, indeed, the *Herald* declared, the children had looked forward to, namely, three addresses: one by each of the two rectors, and by the Rev. Mr. Bailey, of Lowville. And then, at long last, food. The collation out of the way, the children were on their own.

Miller's Grove remained throughout the long excursion era perhaps the most popular of the canal's picnic grounds, though Hayes's Grove, in the northern environs of Hawkinsville, had its champions. Other spots, however, had their moments, as when the Catholic parishes of Forestport and Hawkinsville came by canal to Boonville, where, joined by St. Joseph's, they pick-nicked, as the account had it, at the fair grounds. Kelpy's Grove, just below Port Leyden, was on the list too, and Forestport. Sink Hole Pond was another popular objective. This small body of water was formed by surplus water drawn out of the canal by the waste-weir at lock 70. The depression which it filled touches the State highway a matter of forty rods or so below where the highway crosses the canal.

Up to 1875 the picnic committee hired such boat, or boats, as might be free for the day. In that year, however, A. J. Lee built in Boonville a craft designed to cater especially to the picnic trade—"a jaunty little craft," according to the editor of the *Herald,* "canopied with an awning to protect from 'sun or shower.'" The editor went all out for the "Carrie," for such was her name, adding to his description a plea for business for her. "When completed why," he asked, "would it not be an enjoyable trip to charter it for a feeder trip as far as the Black River at Forestport, and pick-nick upon some island, or some one of the groves that skirt the border of the feeder or river? Choose your committee and appoint a committee for the voyage!"

Not that all this got him anywhere. Not long afterward, with an unwonted respect for starting schedules, the "Carrie," loaded with customers bound for Miller's, shoved off on the stroke of nine. By the time the editor reached the dock the boat was up the feeder, well out of sight. Nothing daunted he went back up-town and engaged a Hulbert House rig to help him catch the party at Hawkinsville. Either the horse was feeling its oats or the driver his corn, for down Main Street both were presently participating in one of the races that were always enlivening the village. Down near Third Street the Hulbert House entry, apparently deciding to break it up, crashed into the hitching post in front of the Goodrich house. Unharmed and optimistic, the editor picked himself out of the debris and started back to get another rig, only to be waylaid by Dan Preston, from over Ava way, who wanted to clear up his subscription account. This matter duly fixed up, after, it seems, an excessive amount of figuring, and another outfit was signed up, he reached Hawkinsville, only to learn that the "Carrie" had long since passed that port. Onward the editor went, in the best traditions of journalism, reaching Miller's just as the clergy had finished the preprandial part of the day's program.

The harrowing experience did not diminish the editor's high regard for the "Carrie." Later on he described with the utmost enthusiasm what one gathers was a notable affair in which the Presbyterian congregation and Sunday-school traveled to Hayes's Grove. "The boat made two trips each way," he declared, "and gave all a chance to go and get out of the glare of an intense July sun for a short season, and enjoy the shadows and cool breezes. Swings, ball, croquet, harsh and gentle sports engaged their attention, until, tired of these, they began to gather in bevies on the green sward or at the tables, to feast upon the goodies teachers and parents had spread for their enjoyment. The moments sped on angel wings, and, weary of pleasure, they gathered their baskets and embarked, homeward bound. The pleasure of the excursion was enhanced by the attendance of the Boonville band."

Picnics always seemed to stir up fine language like that. The paper's Forestport correspondent a few years later gave with an even more ecstatic description of a trip captained by Bill Shanks:

"Beneath a fog, through which, however, the sun's face beamed occasionally with a most delightful and encouraging smile, betokening a fair and glorious day, the boat, whose name the writer cannot recall, commanded by that old salt, Capt. William Shanks, moved proudly out of the pond, thro' the lock, into and down the 'raging canal'—moved majestically along, crowded comfortably with men, women and children. Thus, with music from the Forestport Cornet band, bounding heart and sparkling eyes, the picnic moved onward to its destination. One hundred sat down to the meal and ate, some until they could eat no more."

If the writer could not recall the name of the boat one may be sure it was not the "Ollie," which by that time was in commission. That little ship, a steamer, was owned and skippered by Isaac Scouten. Ike would not at that time have committed the captaincy of his little craft, the smallest in the canal service, to another hand, even if that hand was Bill Shanks, one of the best known boatmen in the last half of the canal's career. Where he found the boat is not known precisely, though she had at one time borne the name of "City of Utica," and it can be assumed that Ike had picked her up in that port. In any event he renamed her the "Ollie," for his young niece, Abe's daughter, and as such she became the sweetheart of the feeder fleet.

The "Ollie" performed all kinds of miscellaneous services. Carrying picnics was only one. On one occasion, when a sudden freshet threatened the dam at Forestport it was the "Ollie" that cleared the docks along the pond of perishable goods and moved them quickly to safety in the calm waters of the Alder Creek Pond. Throughout most of the 1890's she made a regular schedule of daily, and some seasons tri-weekly, trips to Boonville. Leaving her home port in the morning at nine o'clock, she reached Boonville at noon, leaving for the return journey at five.

In her capacity chiefly of charter boat the "Ollie" carried people to all kinds of things—the home people, and people along the feeder, to the Boonville fair; the Forestport Good Templars to union meetings at Boonville, and vice versa; the Hawkinsville theatrical public impartially to Boonville and Forestport runs of popular attractions; Democrats to Democratic rallies, and Republicans to theirs. It was picnics, however, that called out the best in the "Ollie"—as it did in the regular canal boats when they managed to beat the "Ollie" to a turn in this particular trade.

Along the river sector, between Lyons Falls and Carthage, picnics took on grander aspects than those of the kind just described. They were gone on upon a steamer, no less. As when the "F. G. Connell" left Carthage on a "Pic Nic Excursion," as the handbills head-lined it, to Lyons Falls. There was a catch in such jaunts—the cost for long trips, like one in 1866, from Carthage to Lyons Falls, was $1.50 per person. It is only fair to add, however, that in the event just mentioned the Carthage Cornet Band was hired, for the purpose, one supposes, of assisting the participants in enjoying the scenery. Sometimes a Carthage picnic would be held at Smith's Landing, which was Lowville's nearest port. Lowville, for its part, held many of its picnics near Deer River, and as many again at Lyons Falls. One of these picnics was under the management, the advertisements informed the public, of a "Gentlemen's Committee, made up of 15 gents."

A picnic held in 1879 was arranged on an all-out basis. From such facts as are available it seems to have been a fire-company business. Anyhow, the trip started from Carthage, from which place the "Connell" departed at the comparatively unknown hour of five o'clock in the morning. The Lowville and Utica fire departments were aboard, and in a Lyons Falls grove a platform covering a half acre awaited the boat—it had been built, according to the program, "for the amusement of the party"—promoter's language for dancing.

On the steamer "Nellie Sweet," also in 1879, excursionists came from as far away as Lowville to take in the Fourth of July celebration at Lyons Falls. The day's program included at night a grand display of fireworks over the falls, with "rockets, tableaux fire, bengal lights, a lighted balloon, Roman candles, and pin wheels." The balloon added gay and exciting colors to the scene; that cataract, according to the Lowville *Journal and Republican,* "was lit up with blue, red or green, and with fire in all directions over the falls from the candles, etc., making it one of the most magnificent spectacles ever witnessed in Lewis County."

Some of the young folks in Boonville worked out a means more novel than canal boats, of getting to a picnic—row-boats drawn by fleet-footed horses. Of one such picnic in 1880 the Boonville *Herald* had a report that not only paints a pretty picture of the affair, but remains an example of bucolic description at its best. The day's events may be passed over here, until we reach the moment when the sun, setting low in the west, reminded the party that going-home time was at hand. And now let the enraptured reporter take over. "They embarked in their swift-sailing gondolas and proceeded at a high speed down the rapid flowing stream, the heels of the spirited steeds raising large clouds of dust, which floated majestically in the air, making he scenery along the banks of the canal have a more thrilling effect. As they sailed along, congratulating themselves over their safety on the voyage, the tow-line caught on a stake that was near the romantic stream, which shook the occupants of the boats about, causing considerable consternation and alarm among the fair sex, and creating merriment and excitement for the *gallants,* who so heroically enlisted as fresh water sailors and navigators upon the rolling waves and dashing sprays of the Black River Canal feeder. A ride down the feeder, drawn by a horse at full gallop, is one of the most delightful pleasures ever experienced. The boat, riding serenely the silver-capped waves that rolled up and dashed against the high banks of the stream, glides swiftly on, rocking gracefully as the quadruped is urged to its utmost speed."

Rowboats, for that matter, were plentiful in every canal port; so popular was the pastime in Boonville that a group of the younger set got together and built themselves a boat-house near lock 71.

A boat, when it had been painted, and decorated in all kinds of fancy patterns, was in 1880 what a decade later the safety bicycle was to become. No less exciting than the patterns that adorned the boats were the races in which they were always engaging, a form of sport no less popular than the horse-and-buggy races down any Main Street of the time.

Three foreign rowboats showed up in the canal in 1880, when, according to the Rome *Sentinel,* a "novel fleet of boats passed through Rome by canal. The fleet consisted of three rowboats, made of drilling or canvass, stretched over frames and painted. The occupants were W. E. Johnson, C. M. Beckwith, and W. E. Woodward, of Berkeley Divinity School, and E. C. Johnson, of Union College, Schenectady. They left Schenectady bound for the North Woods, *via* the Black River Canal." The flotilla got as far as Port Leyden, from which point they disappeared into the woods, and no more is heard of the expedition, but in the absence of information to the contrary it can be assumed that, tanned and in good brawn, they showed up at their respective schools at the appropriate time, ready to rassle next season with Greek and Latin texts and other phases of the curriculum.

Eye-filling steam yachts ventured now and then into the canal. In 1878 Boonville citizens taking their early morning constitutional along the canal front, gasped at the splendor of a trim craft tied up at the Schuyler C. Thompson dock. Major Barker, of New York, her owner, had come to visit his friend, John Constable, of Constableville. During his sojourn he treated his friends to voyages along the canal, the list of ports reached including Lyons Falls and Forestport. It is clear that the Major was not daunted by the rate of toll charged for pleasure craft, the same being five cents, no less, per mile.

In any list of "fun on and in the canal" activities swimming would have a prominent place. Everybody to whom an adjacent stream was not handier, swam, and also performed his ablutions, in the canal. Especially boys. If the crowds of boys who used it never actually impeded boating traffic it was only because the traffic was not too heavy. In certain of the villages swimming in the canal was frowned upon. Particularly as a Sunday affair. Boonville, as a matter of fact, had an ordinance against Sunday swimming, with a $25-fine to be imposed for each violation. The constables and such were called upon from time to time to enforce it, but nobody ever did, so far as these researches have discovered.

Many are the men still living who as boys learned in the canal levels to swim, getting their first lessons lying on a piece of plank and paddling and splashing with fierce energy. Miss Altsie Wilcox, of Port Leyden, has told the author of looking out upon the canal of a summer's day and seeing afar off what appeared to be giant frogs bearing down upon the village. Speed was not a notable characteristic of this form of locomotion, but in good time the splashing came close enough to reveal itself as a collection of boy-propelled planks.

In many ways wintertime furnished the best recreation on the canal—for there were some who never regarded picknicking as a form of recreation. Sid Rowe was one of these. Sid thought nothing of strapping on his long, curl-nosed skates and whipping up to Forestport and back again. Marathoners like Sid took to the ice in packs and only seldom were content to mingle with those whose idea of a sprint was down to lock 70 and back again.

And if one didn't take to skates he could get out his two-year-old Hambletonian, hitch it to a cutter, and down on the canal hit her up and watch the heel-path breeze go by. This grand sport was made the more exciting when two or more horses were out at the same time, in which case the drivers laid bets on their ponies. Undoubtedly a thriving booking business failed to develop only because everyone who was not in a cutter was out doing figure 8's or was busy putting on his best girl's skates. What may be designated as ice-cuttering lasted practically to the last days of the canal.

The "Ollie," "sweetheart of the feeder fleet," in the canal in Boonville. Photograph by courtesy of Mrs. Henry Thomas, Boonville.

A George Seiter boat, the "George F. Weaver," built by Seiter for Bill Richardson, who also owned and ran at various periods the "Oscar Gorman" and the "Thomas J. McNamara." Photograph by courtesy of Erlo Capron, Boonville.

The Five Combines and old barn. At the tip of the pole furthest to the right is the lock-house at the Lower Threes—Locks 44, 45 and 46. Photograph by courtesy of Roscoe Clark.

State Shop, scow and scow crew force in Boonville. Picture taken from feeder tow-path, looking over canal and railroad toward Main Street. Building still standing in Erwin Park, beside village swimming pool. Photograph by courtesy of Stanley Jackson, Boonville.

Nelson J. Beach, of the town of Watson, in Lewis County, who made the first journey on the Black River Canal. State Canal Commissioner at the time, he had served his town as Supervisor and his County as Chairman of the Board of Supervisors and as Assemblyman, and his Senatorial district as State Senator. After a term as State Canal Appraiser, and then as Resident Engineer of the Eastern Division of the Erie Canal, he was made General Superintendent, and later Vice President, of the Hudson River Railroad. After each of these assignments, he would return to his farm to pick up where he had left off, but never was allowed to remain for long. He died in 1876 at the age of seventy-six years. Photograph by courtesy of Mrs. Georgiana Beach Bowen, Lowville.

The steamer "Edith M. Van Amber," owned by Van Amber Brothers, with a tow of lumber-laden boats emerging from the Otter Creek lock below Glenfield. Photograph, taken about 1890, loaned by courtesy of Mrs. Spencer E. Burdick, Glenfield.

Peter Phillips, Superintendent of Section 1 of the Black River Canal from 1883 to 1889. Upon his retirement it was said of him, "He was a practical canal man, and the present splendid condition of the section under his recent control is a tribute to his efficiency." In the canal season of 1886 not a single complaint, by boatmen or others, was entered against management. Photograph loaned by Mrs. Grace Phillips Gibbs, Kalamazoo, Michigan.

Lock No. 71, in Boonville, marked Canal's end, as of 1850. The buildings seen in the foreground were constructed subsequent to that year. Photograph by courtesy of Ray Schweinsberg, Boonville

A canal boat, the "Hiram and Wilber" owned by the Van Amber brothers, lumber manufacturers, of the town of New Bremen, at the southern end of the canal in Boonville. On the tow path, the tow-team can be made out. The first structure at the left is what with the coming of the canal was the Gilbert & Holdridge foundry. In the background, at the head of the canal, where it veers to the left was the canal shop and barn, later to be used as a paint shop. Photograph by courtesy of Karl Traxel, Boonville.

Above, the "Dudley Capron," built by Frank Seiter at Boonville, Sidney Phillips, Captain, unloading sand in Syracuse. On the boat are Sidney and his mother. Left, sluice at Lock 35 along the Lansing Kill. Photographs by courtesy of Erlo Capron, Boonville.

The "Annie Laurie," one of the best known boats of the Black River fleet, owned and run by Arthur Lane, of Port Leyden. "Archie," as he was known up and down the canal, is seen in the center of the group; to the left is Mrs. Lane; to the right is son Walter, who himself in the later days of the canal boated with his father. Photograph loaned by courtesy of Walter Lane, still of Port Leyden.

Remains of the lock at the mouth of Otter Creek on the Black River Improvement, looking down the river. At the far end of the lock the west gate may be discerned, its planking gone, but hanging on the original irons. Photograph made by Alice E. O'Donnell in the autumn of 1948.

The "Timothy Curtin," owned and run by Michael Grems, with his wife, Mrs. Bertha M. Grems, seated on the cabin. The boat was the first boat to be locked through the new combines at the Delta Dam in 1912. Photograph by courtesy of Miss Nellie Wetmore, Lowville.

Part of a tow of lumber-laden boats emerging south-bound from the lock at the mouth of Otter Creek, two miles below Glenfield. Photograph by courtesy of Mrs. Spencer E. Burdick, Glenfield.

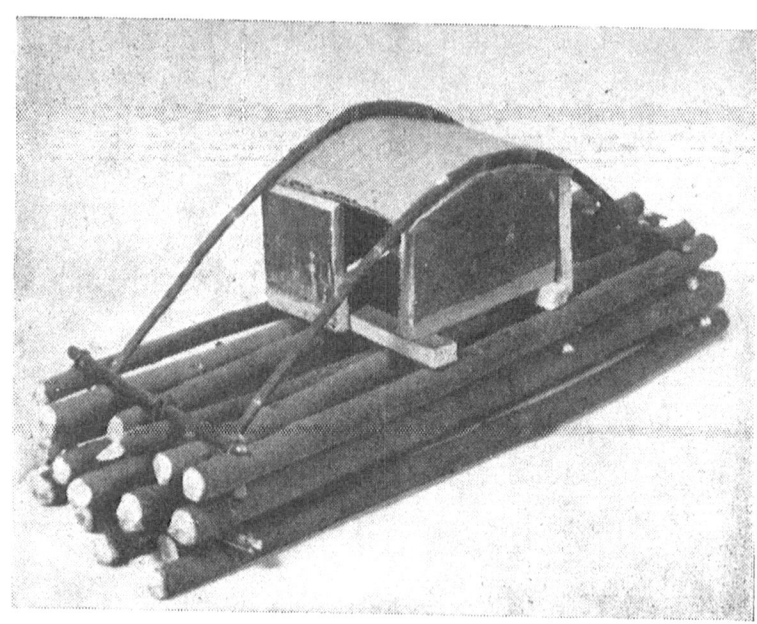

Model of a timber "crib" made by Roscoe Clark. The last crib in each "tow" was the "shanty crib," so-called because on it was a shanty, divided into two sections, one for the crew and the other for the horses. South-bound boats meeting north-bound boats held to the tow-path side of the canal, which meant that the tow-lines of north-bound boats would pass over the cribs of timber. To facilitate passage of the lines over the shanty was the purpose of the two arching "shanty poles."

Looking north from Lock 35 towards the Five Combines. Photograph by courtesy of Mrs. Olney Clark.

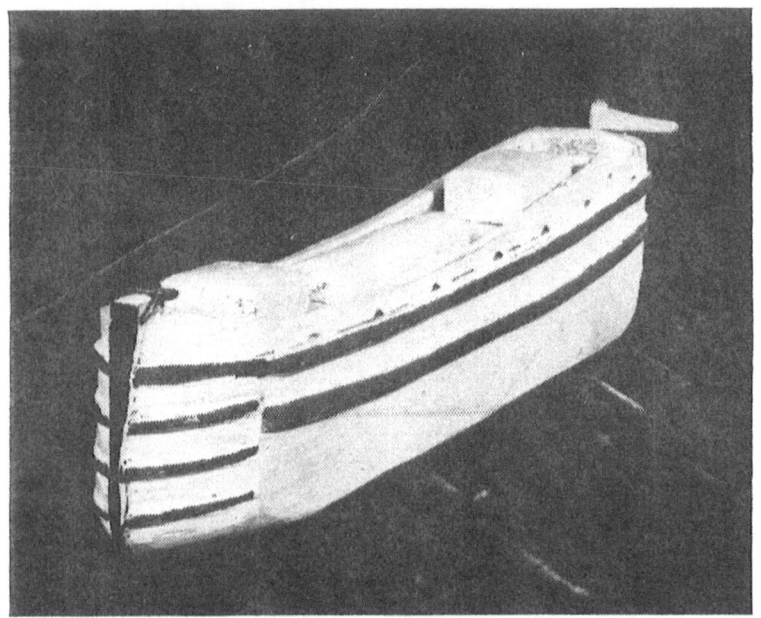

Model of a "laker" type of canal boat, distinguished chiefly by the rounded nose. Made by Roscoe Clark, veteran locktender of the Five Combines.

Looking up the canal toward Dunn Brook, seen in the background. In the background the roof of the Ed Keyes store shows above the trees. Photograph loaned by William H. Rogers, Boonville, for many years tender of Lock No. 30.

Henry Abbey's mill at Abbeyville on Independence Creek. Photograph by courtesy of Charles Abbey, Glenfield.

Second Basselin mill, built to replace previous structure destroyed in 1899. This second mill burned in 1909. Photograph by courtesy of Ben Bachman, Naumburg.

The second Van Amber mill, which burned in 1900, the fire caused by sparks from a passing Basselin steamer. Photograph by courtesy of Mrs. Harriet Van Amber Meeker, Lowville.

The Van Amber establishment as it appeared in the early 1890's, showing boat-building shop at left, and at right of center the mill. When boats were launched at the shop they were moved to a dry-dock, a half-mile down stream, for finishing. The structure of a boat nearing completion may be observed below the shop. The previous mill was destroyed by fire in 1883. Photograph by Mrs. Harriet Van Amber Meeker, Lowville.

The celebrated "Pasenger's," at Bush's Landing, in the town of Watson, Lewis County. On the balcony railing may be seen the stuffed skin of a panther, shot over at Number Four by Bord Edwards. At right on the balcony is Mrs. Pasenger, Jack either being away for the day, or, more likely, down showing the photographer how to do it. From an old photograph by courtesy of Miss Nellie Wetmore, Lowville.

The Wellington Brown Hotel in the town of Greig, near Glenfield. The picture was made after the establishment had been taken over by George Dekin, who with his family are shown in the picture. Photograph kindly loaned by Mrs. D. P. Carey, Glenfield.

The Black River steamer, "J. F. McCoy," owned by Van Amber Brothers, used for towing canal boats in the river section, tied up at Lyons Falls. Photograph, taken in the 1880's, used by courtesy of Mrs. Harriet Van Amber Meeker, Lowville.

Repairing bridge over Black River at Glenfield. About 1890. Photograph by courtesy of Mrs. Spencer E. Burdick, Glenfield.

The "Grover Cleveland," a T. B. Basselin boat, entering lock 7, which was tended by Fred N. Miller, and is now covered by the Delta Reservoir. Photograph by courtesy of Fred N. Miller, Boonville.

The "Harrison and Kenneth," owned by Tom Shanks, of Forestport. A Durhamville "laker" named for his two sons — distinguished as a "laker" because of its rounded prow. Photograph loaned by courtesy of Erlo Capron, Boonville.

The level between the "Lower Threes," looking toward the Five Combines. Photograph by courtesy of John Shepard, Boonville.

The dam built on the Black River at the mouth of Otter Creek. Just beyond the fringe of trees in the background was the by-passing lock. Photograph, taken about 1890, loaned by courtesy of Mrs. Spencer E. Burdick, Glenfield.

Culvert No. 4, which carried the creek from the Clark gorge under the canal near Lock No. 36, a half mile below the Five Combines. The canal as seen in the picture is a good illustration of how the canal had not only to be dug but in many places built up. Photograph, made in 1948, by courtesy of Mrs. Olney Clark.

Baker's Falls, formed of water from the waste-weir at lock 70. Photograph by courtesy of John Shepard, Boonville.

8

Early Boats and Boatmen

Up to the completion of the lock and dam at Bush's Landing in 1866 navigation was pretty much in the hands of "foreign" boats. Meaning boats moving into the Black River from other canals. Records of the period are scanty. Black River boats were never registered; the registry of clearances at the toll offices in Boonville and Lyons Falls would give the entire story were they complete: few of the books remain, however, and upon the scanty information which they give history must rely for its data. To them the present writer has turned for much of the information offered in this chapter.

Boats coming into the Black River country hailed from every corner of the State and from every canal. The "Bill Camel" came from Elmira by way of the Chemung Canal, and the "John Cramer" from the Champlain; the "Oneida," owned in New Hartford, came in from the Chenango Canal, the "Two Brothers" from the Oswego, the "Keystone" and the "A. L. Burr" from the Caneadea country on the Genesee Valley canal, and the "A. Newell, Jr." from the Delaware and Hudson Canal down in Ulster County.

The Black River fleet, however, was coming along. That proud boat, the "Major Anderson of Fort Sumpter," named and put into commission just after the outbreak of the Civil War, and in memory of the Major's heroic defense of that post, was going strong. The boat's name, it should be recorded, created no little havoc in the records of the Boonville toll office. The clearance records had a column for the name of the boat being cleared, and next to it a column for the place from which it hailed. All entries of the "Major" gave, in the first column, the words "Major Anderson," and, in the adjoining "from" column "Fort Sumpter."

In service in 1866 was the "A. M. Collier," of Forestport, A. H. Kilmer the master. And the "Northern Light," of Boonville, Captain Philo Kane, that year carrying mostly lumber from Sugar River to Schenectady and Albany. That year the "Maid of Judah," a famed boat which, with its skipper, Captain Bill Roe, of Westernville, is remembered vividly to this day by old boatmen still living, was also in lumber. Bill himself is best remembered, perhaps, for the great physical power of his wife Emma—the name undoubtedly will be news to one and all of the old-timers, since she was known universally as "Em." She boated throughout the years with Bill, and while she never sought a battle, yet in defense of her husband, at such times as strangers among boatmen thought to take liberties with him, she leaped to his defense, landing upon the invaders with left-hooks and such with the utmost abandon. An oft-told story is of how, hearing a disturbance upon the tow-path, she climbed out of the cabin, jumped to the bank with a single leap, dropped a haymaker upon the jaw of the surprised intruder, and retired calmly to the cabin to go on mixing dough for the supper biscuits as though nothing had happened. From all accounts this episode was typical of scores of such meritorious endeavors. She is remembered with the utmost respect by all old-timers among the boatmen, who testify to the immaculate condition of the cabin of the "Maid of Judah," with the bright chintz curtains and the geraniums and other flowers that adorned its windows and to the constant devotion with which the Roes took part in divine services when they were home in Westernville.

From Boonville hailed the "Highlander," Captain A. A. Smith, carrying lumber from Hawkinsville, and the "G. O. Bridgman," T. Shover, captain. This latter craft was named for the young man of the same name who at the time was clerk to the Collector of Tolls in Boonville, and whose father had been Superintendent of the canal and now was of the firm of Bridgman Brothers, boat repairers, with a shop at the foot of Second Street. The brothers had recently attracted a good deal of attention among canal men by reason of an apparatus which they had devised for raising boats onto the canal bank. The device consisted of a pair of rails extending down into the canal basin; upon them moved an "inclined carriage," as the partners explained it, the carriage, with boat aboard, being raised by capstan and chains anchored upon the bank. "G. O." later was to be the first superintendent of the village's electric light plant.

A number of boats were by this time, 1866, running out of Carthage, among them the "M. M. Carter," Captain J. Shaver, carrying pig-iron and general merchandise to ports along the river and canal, and the "A. P. Gilbert," Captain D. Rhino, carrying this year several loads of leather out of Carthage. Then there was "Ella's Sister," Captain G. Denslow, in the lumber trade—one cargo was made up entirely of barrel staves and heading, consigned to Troy; the "S. S. Hoyt," Captain George Davis, engaged in the port-to-port pick-up trade, and the "C. Dodge," Captain S. Isbell, lumber.

Besides the "Maid of Judah" other boats are listed as hailing from Westernville. The "Two Sisters" is one, given on one trip as in command of Captain Bray, and on another as in charge of "Captain Healey." Another Westernville boat was the "Topsy," Captain Bronson, and the "M. Brayton," Captain John Sherman.

From Hawkinsville came the "Frank Brown," captained variously by Hank Sherman and O. Clark, from home base carrying wood mostly, and from Port Leyden lumber. From Forestport one of the best loved men in the canal service, Isaac Scouten, who had lost an arm in the late war, ran the "C. Willover," in the Black River-Albany lumber trade. From the same port came the "A. G. Seeley," Captain R. Wormwood; the "Fred Griffin," Captain J. C. Hovey, the "A. M. Collier," and Tom Nightingale with his "J. Carr."

Two boats were cleared in Boonville as belonging to North Western, the "Charles Park," Captain H. Philkins, and the "Henry and Helen," Captain Marcus Van Buskirk. This latter boat was built in North Western by Jerome V. Gue, whose boat-building activities have been noted in an earlier chapter. In a later year the "Henry and Helen" was captained by Allen Phillips, one of a family of brothers whose physical exploits rivalled those of Em Roe, and one of whom, Peter Phillips, was to become Superintendent of the canal. More of these remarkable brothers will appear in its place.

From Lyons Falls is listed the "Florence," Captain H. D. Hemstreet, which during the year carried several cargoes of wood from Port Leyden to Utica. For its own contribution to the list of Black River boats, Port Leyden had the "James C. Pullman," Captain Charles Sherman, engaged in the lumber trade.

Of all the boats named only the "Maid of Judah" and the "Henry and Helen" remained a decade later in service—at any rate under the same names. Since boats were not registered, and they frequently were changing hands, and a transfer often involved a change in name, the history of any given boat can seldom be followed from data yielded by the meager official reports.

Up to this year of 1866 the tonnage cleared through the canal had been steadily increasing. Since 1851, the first full year of navigation, with a total of 25,320 tons, the volume had grown until 1865 could report 73,317 tons. The next year, reacting to the impact of the full operation of the entire canal north to Carthage, turned in a total of 85,908 tons. Since the average load of a Black River boat was around fifty tons, it is clear that the boats owned along the canal were inadequate to the handling of the traffic. What would have been an acute shortage of boats was obviated by the large number of foreign boats. Overnight, however, boat-building on the canal took on all the aspects of a new industry. Gue had already started at North Western, and presently excellent boats were also being turned out in Boonville, and a little later at Hawkinsville, which over the years turned out more boats perhaps than any other port. With a lesser boat production record came Lyons Falls, New Bremen, and Forestport.

As the home fleet grew the number of boats coming into the Black River country from other canals dropped away. The flood of lumber, timber, wood and other timber products was already under way, in 1866 amounting, for lumber and scantling, to a total of 29,157,124 board feet; timber, 134,751 cubic feet, and wood, 9,503 cords. The figure for lumber as just given represented an increase over the previous year of 5,680,312 feet, and was under 1867 by 1,009,821 feet. To what dimensions the lumber business was growing may be seen in the fact that Lyons Falls alone, in 1869, shipped 15,254,258 feet, and the next year 17,267,620 feet.

Meantime the shipping of spruce for spars and piles, the latter used for building the piers along the New York waterfront, was growing. These giant sticks of timber, limited in length only by that of the locks, was in 1866 an infant industry, but, shipped in cribs of as many as

15 pieces, with five and six cribs in a tow, spars and piles were to reach, ten years later, the astonishing volume of more than a hundred thousand pieces a year.

It was this lumber and timber business, with its glowing prospects, that Black River men wanted to get in on. All of which meant boats. And boats they built, and to the extent that the growing fleet not only took care of the increases in tonnage, but also gained what was virtually a monopoly of the Black River carrying business.

9

Family Business

By 1875 the pattern that began to emerge from early canal operation and canal traffic as they stood in 1866 had completed itself. Traffic was mainly by boats built, owned and run by Black River men. Many of the families that had gone into boating remained, through three or four generations, to the end. Not that the business was not open always to newcomers. Anybody who had from three to five hundred dollars could pick up an old boat, and, finding a mill man with high piles of lumber to be moved, or a potato buyer with a deep yearning to get his spuds to New York, he would be in canal navigation.

If a man had from twelve to fifteen hundred dollars, of course, he could go to Sam Ferguson in Boonville, Frank Seiter in Hawkinsville—to any of the builders of boats, and get a brand new number just out of the dry dock, or even have one made to his order. Once in, however, the chances were that he was in the business for keeps. The thing would get in his blood; he had become one of a group of men bound together by a kind of freemasonry of the canal levels; there would be seasons when he couldn't make a dime; death was always just around the corner, ready to strike out at a boatman, and, if they traveled with him, at members of his family; during eight months of the year he was a stranger to his home, and unless they were with him, to his family as well.

The hazards and the hardships of a boatman's life, the stark tragedies and the humors, grim and ungrim, were the ingredients of what would make up in a man the canal spirit—a way of looking at things, a way of taking fun and hard, sweaty work as they came, a stoical acceptance of personal grief—that would give color and flavor to his life ever after.

Into the pattern also had come boating families, made up often of men who carried goods which they themselves produced. Meaning lumber manufacturers, like the Beach brothers, Andrew and Ralph, and the Van Ambers, Watson and Henry. Mostly these families built their own boats, sometimes requiring a fleet of them, which would require the hiring of captains to run them.

The Beaches built at Bush's Landing four boats of which record exists. They were the "Ada P. Beach," the "Andrew J. Beach," the "Jennie Beach," and the "R. Beach Jr. & Sons." A boat variously showing up in the records as the "R. J. Williams" and the "R. Williams" was also built at Bush's Landing, and may have been a Beach boat, although Henry Wetmore, of Lowville, in conversation with the author, gave Robert Williams as the builder. Nephews of Nelson J. Beach, already noted in these chapters, the brothers had vast timber holdings in back of the Landing, and proceeded to saw, and themselves to haul, their lumber to market. The boating problem was not a major one, there being enough sons in the two families to man the boats. Andrew himself was an inveterate and able boatman, and usually carried his family, including such of the male members as were not old enough to run boats.

In course of time the "Ada" was sold to Jerome Salmons, from up Otter Creek way, but as late as 1883 Boonville clearance records give Andrew still as the captain. The "Andrew" likewise was owned by Marcus Van Buskirk, of North Western and William Norton. The "Jennie Beach," when its turn came to be sold, was taken over by George Kelley, and thereafter, at least under that name, disappears from the records. Confusion surrounds the name of the "A. J. Beach Jr. and Sons." In this form the name has been given by old boatmen with exceptional memories; as the "R. Beach & Sons," John L. Beach captain, however, it was cleared in Boonville in 1875, and in 1876 as the "R. Beach, Jr. and Sons," Patrick Dugan its captain. In conversations with old boatmen an "Isaac P. Beach" has appeared, but the name shows up in no clearance records, and is here set down for the record.

Similar to the Beach story is that of the Van Amber boats, owned by two of the Van Amber brothers, Watson and Henry, blue-eyed, red-bearded, black-haired, both of them. They were big-scale lumber producers, their mill, owned by Watson, situated a mile up-river from the mouth of the Beaver, in the town of New Bremen. The Van Ambers, like the Beach brothers, built their own boats, and ran some of them. These researches have identified ten of them, all reported to the author with what he accepts as reliable authority, and some having official clearance records.

There was the "B. Van Amber," afterwards bought by Harvey Boyce, of Forestport, who, with his brothers, Jim, Joe and Charley, made up a considerable boating family. Harvey was afterwards captain of the "Emory Allen," owned still later by Henry Brockert. Another Van Amber boat was the "B. G. Kimball," later owned by George Ferguson, and the "Estella F. Tuttle," its captain Warren Tuttle. Tuttle later owned the "Cold Wave," which Ben Cooper, of Port Leyden, recalls well, having run on the boat with Warren.

The most famous of the Van Amber boats was the "Hattie and Lizzie," the Hattie thus celebrated being the daughter of Watson, while Lizzie was the daughter of Henry. Both cousins, it is the author's happy privilege to state, still live near the scene of their childhood, recalling the days when their father's boats set out, loaded with sweet, fresh sawn lumber, for Albany or New York. So commonly was the boat's name abbreviated along the canal that it must have come always as a surprise to such boatmen as might learn that the real and official name was not the "Hat and Liz." The earliest official record in surviving clearance papers is of 1883, when the boat was cleared in Boonvile, carrying 22,000 feet each of hemlock and spruce lumber, the captain George Eddy, with whom running the "Hattie and Lizzy" seems to have been a kind of career. The boat had a long life, and an uneventful one, the only recorded accident occurring on May 18, 1898, when it sank in the Forestport feeder, holding up navigation for twenty-four hours before it could be raised and set on its way, to the utter disgust, no doubt, of the tow team.

Other contributions to the canal fleet were the "John Donley," run by John Hanley, and at a later period by his brother Mike; the "Lillie M. Nichols" and the "Mable Ervin," each having William Nichols as its captain; that pair of sister boats, the "Rousseau" and the "Hiram and Wilbur," and the "Nellie Van Amber," the latter owned and run later on by Charley Gleasman, of Forestport. On a day in June, 1894, the "Nellie," striking a rock in the feeder, sank, in the feeder's normal four feet of water. Thirty hours were required to lighten the load to the point where it could be got down to North Western for repairs.

T. B. Basselin, the river's largest producer of lumber, built, first at his Beaver Falls mills, and afterwards at the huge Castorland mill which in 1885 he built on the west bank of the Black River, opposite the mouth of the Beaver, a number of boats to get his lumber to market. Those built at the Castorland mill could qualify as "Seiter boats," inasmuch as the work of designing and construction was in the hands of the capable Lewis Seiter, of Boonville and Hawkinsville, whom Basselin had engaged as foreman of the works.

Three boats known to have been built at Beaver Falls, and which therefore would be in the Basselin list, bore the names of the "Eva May," "The City of Venice," and the "Ezra Benedict." The known Castorland craft, for identification of which the author is indebted to Roscoe Clark, who remembers them clearly, were the "Adlai Stevenson," Matt Perkins captain, and at a later day owned and run by Dave Golden; the "Francis S. Norton," owned at various times by Leo Remp and Jim Johnson; the "Grover Cleveland," John Plato, captain; the "Zena E. Harvey," named for the granddaughter of the captain, Henry Harvey; the "M. W. Holt," Dell Satterly, captain, and the "Peter J. Rohr," its captain Jim Johnson, of Port Leyden, with whose death in 1914 a long and honorable canal career came to an end.

In addition to running the "Rohr" Jim had owned the "Charter Oak," and other boats, had been captain of the State scow, and in the river section of the canal had been in charge of channel dredging. Along the Watson sector he is remembered with affection by old canal men who were associated with him; one such, outlining to the author his river activities, mentioned with pride that he had once worked for "Jimmy Johnson." He had served honorably as a cavalryman in the War of the Rebellion, and in the course of an obituary article the editor of the Boonville *Herald* could say with all truth, "He was jovial by nature and had a good word to say of all, whether friend or foe. Under a rugged exterior he carried all the sterling qualities of a noble and honorable manhood. His associations with neighbor and friend were expressive of the cordial feelings he entertained toward all, which won him a place in their hearts never dislodged."

Here may be included, too, a report of a family of brothers who for many years enlivened the scene and left an indelible impression upon the history of the canal. They were not shippers, like the Beach and Van Amber families, but were carriers on a very considerable scale, and also played an important part in the canal's administrative and maintenance operations. When the last brother had retired from the canal, the story ended then and there. Referring to the Phillipses—Pete, Allen and Alex, who lived at various times, one or more of them, along the Lansing Kill, in Boonville, or in Hawkinsville. A fourth brother, Jim, lived in Lyons Falls, and had fewer canal associations. Not only were the trio excellent boatmen, but they had the temperament of bear-cats, and the fistic cunning and sheer physical power to back it up. Old boatmen who remember them well, declare that they never started a fight, just for the sake of fighting; bullies they were not, but on the other hand nobody who ever mistakenly tried to bully them but regretted, thirty seconds later, that they had undertaken what presently turned into a misadventure. At breaking up brawls they were particularly expert. Coming upon an over-size brawl, something worthy of their talents, they waded in with well deposited lefts and rights and a moment later came out to lay the brawlers on the ground in neat and orderly rows.

By all reliable accounts, Allen of the three was most competent in this field of endeavor. And the top performance of his career was to occur, not on home grounds, but down on the Erie while snubbed up in the vicinity of Sprakers. The affair, so far as the Erie men were concerned, was undoubtedly the result of bad feeling left over from some previous encounter with Black River men. Be that as it may, a gang of Buffalo boatmen, set on strewing the remains of some, any, Black River boatman around as a new feature of the Mohawk scenery, pounced upon Allen. The story does not picture Allen as taking many of their insults, for he was presently in the midst of as ferocious a battle, judged by such few details as remain, as even a Black River connoisseur of tow-path battles could wish. He faced elements of two boat crews, but in he bored, with the old Phillips spirit, and was making good headway when someone came at him with a windlass bar that caught an arm and broke it. Alex now got really mad and when the battle was ended the wounded Eriemen were lying all over the place, with Allen having the field all to himself.

Boats which the brothers owned and ran, and which, owned by others but named for them, make up a considerable list. Pete had the "Charley Thorp" in 1875, and later the "Emma Jane." Allen's boats included the "Henry and Helen" in 1875, and subsequently the "Hattie," named for his daughter; the "Peter Phillips" for his brother; the "S. O'Connor," the "Theodore Seeber," the "Frank and Ollie," and the "George R. Ainsworth." He had a practice of buying new boats, running them awhile, only to sell them and again buy new ones. His hobby was swell harness turnouts. On such occasions as he came along the Hawkinsville avenues, behind a smartly matched pair of grays, harness and harness accessories shining, and the buggy all new and glittery, the populace was stopped in its tracks.

Boats owned by Alex included in 1876 the "Fannie L. Seiter" and the "Katie Seiter," both boats built (as were most of the Phillips' boats, for that matter) by Frank Seiter in his flourishing plant just up-river from Hawkinsville. Later he had the "Arthur Phillips," and the "Herbie and Hattie."

The above boats are but a partial list, compiled from clearance records and interviews with boatmen whose memories go back to the Phillips era. One season alone, 1893, Allen and Alex between them were reported as having fourteen boats tied up at Forestport. That was the year when the famous panic of the 1890's struck most savagely, leaving Black River shippers and boatmen with the dullest season the canal had ever known.

It was the year also of the Forestport sea-monster scare, when a citizen, writing of the occurrence, reported that "a strange animal, or fish of huge dimensions, was seen in the pond making a great disturbance in the water. One said the animal had a head like a horse, and the body of a fish, another that it spouted water and had every appearance of a whale; one that it was undoubtedly a sea-lion, and still another that a huge water-serpent was making its home in the pond.

"Crowds visited the spot and watched impatiently for the strange appearance. After long waiting the water began to move, slowly at first, then more rapidly, and a great brown body appeared above the surface, floated for a moment, then sank beneath the water." Many of the more curious "built a bon-fire on shore and watched all night to see fish or serpent come forth, but up to the present time the strange moving of the water is still unexplained."

Alex, who was in town looking after his boats, had an explanation, to the effect that it might be a body of sawdust, raised by accumulated gases to the surface, and then of its own weight sinking again. That of course, was a prosaic kind of theory, and the people waited on, but to this day the whole business is a mystery.

Pete, leader of the Phillips clan, rose to the height of his career when in 1883 he became Superintendent of the canal, continuing until 1889, after hanging up a record, not only for the long period of his service in this office, but for efficiency of administration, in both respects a record that was to stand as an all-time high. The work he did as Superintendent will be reviewed in a later chapter. It remains here to note that after retirement from the canal in 1889 he interested himself in a number of enterprises, among them the proprietorship of the Railroad House in Boonville. He died in 1893, at the age of fifty-six, and a tribute written at the time could say what all who knew him had in their minds: "The death of Peter Phillips is a sad and crushing blow to the members of his family and the thousands of people who knew and loved this generous, kind-hearted and public-spirited citizen. Today he is dead, and many a silent tear will be shed upon the casket of him who was everybody's friend, who shared his means with anyone who was in want, who was generous to a fault, and who had not an enemy in the world."

10

More About Boat Building

Boats built along the canal, at and south of Lyons Falls, came mostly from a half dozen plants. A number of boats, when the Chenango and other laterals were abandoned by the State in 1873, were brought into the Black River service. Moreover, a few Black River men went to Durhamville for their boats, built there at the Doran establishment. Such a one was Tom Shanks, of Forestport. Tom had practically an all-Durhamville fleet.

The Black River Canal, however, depended mainly upon boats made by its own people, and of these Seiter-built boats heads the list in point of numbers. Especially boats built by Frank Seiter, who was a kind of canal-boat Henry Ford. When it was all over Frank claimed that he had built some forty boats, and the claim was probably on the modest side.

Frank was but one of four Seiter brothers who at one time or another were in the business. At periods he was associated with Lewis; Lewis, however, was for several years at Castorland, building boats for T. B. Basselin, and for the greater part each operated alone. The scene of Frank's first efforts was a parcel of land which he bought a mile and a half up-river from Hawkinsville. Somewhat later he moved to Boonville, where he operated a mill on the canal at the foot of second Street. When fire destroyed the plant he went back to Hawkinsville and operated on the old family homestead adjoining his former site. Most of the boats owned by Peter Phillips and his brothers were built by him.

Lewis was associated with Frank before his sojourn at Castorland, but on his return to home grounds he went in chiefly for house-building, and became to village homes in Boonville what Frank was to canal boats.

John, a third brother, took a turn at boat building, but had time for but two or three when he removed to West Leyden, where canal boats were not particularly in vogue. George, who had bought the old Samuel Johnson tannery property on Mill Creek, below the canal, put out a small number of boats, but he gave himself chiefly to his thriving business in lumber, logs being brought in for him by the feeder from Forestport and spots along the way, and timber hauled in from over beyond Black River.

Sam Ferguson, of the firm of Breen and Ferguson in Boonville, ran Frank Seiter a close construction second. In point of time he was the first of the Boonville builders. The first record of the concern's activities is found in 1858, when a new dry dock was built. Information thereafter is scant until 1861 when the "Major Anderson of Fort Sumpter" reminded the public that the Civil War was now on, while the next recorded event was the building in 1863 of a store, where groceries and boating supplies would be "sold for the accommodation of boatmen"—meaning, of course, that boatmen would be saved the wearisome trip up to Main Street.

When the Utica and Black River Rail Road extended its line on north in 1867, in the village following a new route along the canal, the depot brought new importance to the Breen and Ferguson location nearby, and the partners built a new store, with two floors no less, where not only boating supplies and foodstuffs, but liquid refreshments, were sold. This was in 1868.

Something went, or was going, wrong in the plans, for the two men dissolved partnership the following year. Ferguson continued to operate on his own, but in 1874 Robert H. Roberts, who was conducting a lively shipping business, took over the dry-dock. Roberts hailed originally from Constableville, where his parents, coming to America in 1839, their two-year old son with them, had settled. As a young man Robert worked in Boonville as a carpenter, leaving shortly for the Pennsylvania oil fields. After a few years among the derricks he returned to Boonville and took over the Ferguson dry-dock. Sam, however, continued to use the facilities in his own operations. Roberts was making a few boats, but his arrangement with Ferguson was a break for him, for he was to be in Albany for several years, first as Assemblyman for the Fourth District, and then as Senator.

Finally in 1880 Ferguson himself took a vacation from building and accepted the strenuous post of Superintendent of the Canal—

"strenuous" because beginning with this year the entire canal was consolidated into one section. He held the office for three years, relinquishing the superintendency to Peter Phillips. Ferguson brought a new efficiency into administration of the canal, and it was this pleasant situation that Peter inherited, to relinquish the post six years later with the physical condition of the canal and the morale of the boatmen and shippers at a new high.

H. J. Goodwin, whose "forwarding business" was established in Boonville in the old Joslin ware house near the bottom of Second Street, also built boats back in these early days—not many of them, but, traditions all have it, all excellent craft. Only one can be identified by name, the "C. W. Havens," Henry Williams captain, built in 1862. Havens, it would seem, from such records as exist, was the man in charge of the works, and his name was given to the boat as a happy gesture by the skipper.

C. W. Thorp was another canal-side shipper who took a turn at boat building. None of his boats can be identified with certainty; a number of boats can be put down as built by Ferguson *or* Thorp, and that is all. This was in the late 1870's, and Ferguson, who was a kind of free-lance at this period, made his own boats in the Roberts yard, and may well be presumed to have built for Thorp, or perhaps used the Thorp facilities for making some of his own boats.

The boat repair and building operations of Bridgman Brothers has been noted in a previous chapter. Only one of their boats can be identified, the "Major C. A. Riggs," built for P. B. Lane, of Forestport, for the grain carrying trade. The date was 1866. The Major of the name was one of Boonville's most popular young men, with a reputation for humor and wit.

So much for Boonville. At Lyons Falls Jesse Irons was carrying on at his dry-dock down at the combined 108 and 109 locks, which let the boats down into the river. Names of four boats built by Irons remain: the "Frank and Lib," Jerry Holdridge captain, in 1883; the "Fannie Markham"; the "Lottie Corser," Captain Leman Corser, and the "Matt and Jess," Henry Cunningham captain.

At Forestport the thriving lumber and saw-mill enterprise of Syphert and Harrig was making its contribution to the Black River fleet, with close to a dozen boats that can be identified. Both William E. Syphert and Albert Harrig were Forestport born, and both had started out as canal hands. Harrig at twelve years broke in as driver; at sixteen he was promoted to steersman, and at twenty he had his own boat. Syphert started even earlier—as driver at eleven, shortly advancing to steersman, but at eighteen leaving to work in the lumber woods. Many old North Country men will remember him best perhaps as the brother of Gus Syphert, the great Brown's Tract guide.

At North Western Jerome V. Gue turned out a considerable list of good boats at the scene of his many canal-side business activities in the southern part of the village, boats that will show up from time to time in this history. In Rome L. B. Sherman turned out for the Black River fleet the "Charter Oak," already mentioned as owned at one time by Jim Johnson, of Port Leyden, and afterward at various periods by Hank Gallagher and Charley Boyce; the "Marie Hyde," launched in 1882, and the "Henry Fitzgerald," owned by Tom Shanks. The "Henry Fitzgerald" in 1901 was in an accident that had the maintenance people wondering if maybe it wouldn't have been better if Shanks had stuck to Durhamville boats.

The scene of the trouble was lock 35, which had just undergone repairs so extensive and thorough as to bring a statement from Bert Calen, the Superintendent, that it would last as long as the canal itself, if not longer. The "Fitzgerald" has just got into the lock when suddenly the walls gave way and when it was all over the boat was left resting with its stern in the lock and the bow out on the berme bank. Traffic was held up for thirty-six hours. Shanks declared that had it been a Doran boat the whole lock would have been blasted over into the Lansing Kill.

A couple of men on the side-lines also joined the fraternity of builders. Meaning Michael Hogan, of the town of Western, who built and ran the "Nellie R. Hogan"; Olney Sherman, of Elm Flats, below Boonville, with the "Safe Return," and Vol Pixley, who lived at what is now known as Pixley Falls. An accomplished carpenter, in 1881 Vol built two boats, which were bought by Lewis County gentlemen for the downriver lumber trade. The boats were the "Col. J. Van Woert," Captain Frank Ashcroft, and in 1883 listed in clearance records as hailing from Greig, and the "Willsie Alexander," Captain Andrew Alexander, also listed as from Greig. The "Colonel" later was owned by Arch Plato, hauling lumber for Denton and Waterbury, of Forestport.

Vol's canal activities included the shipping of tows of timber to Albany and the cutting and hauling of wood, of which his lands yielded considerable quantities, and which he sold in Utica and other markets. Vol's brother, Jay, was also a great wood man, and in the business he had the incomparable assistance of Dick Warren, a giant of a man, standing six feet and more, and famed along the canal as a consummate chopper of wood.

By such accounts as remain Dick's technique was no great shakes, but what it lacked in artistry it made up in volume of output. When a day's work was finished high up on the Lansing Kill hills Dick left behind him a cordage that, while creating no great enthusiasm in the minds of other choppers, yet was a delight to any employer who paid him by the day. His like must have been in the mind of a man down at Hillside whose experience with woodchoppers had left him a confirmed cynic. Writing a letter to the editor of the Boonville *Herald* he importuned him to "please inform those persons who may pass through your village seeking work that a limited number of *good choppers* can find employment in this vicinity chopping wood. That class of choppers, however, who tire out soon and are obliged to visit the tavern to *recruit their strength are not needed.* We have enough here already who perform great exploits after taking a few '*nips*' and boast of putting up three or four cords of wood a day. But their labors never amount to anything, for it is all performed by a good fire in the bar-room." What the man clearly needed was Dick, one of whose favorite remarks was that he was "purty well took up with choppin'."

The secret of Dick's power, the Dunn Brook employer would have been glad to know, was tea. On a job he always took a pot of the substance into the woods, and quaffed generous helpings at meals, which led Mrs. Pixley one time to say that he consumed more tea than all the rest of the family put together.

The trouble with Dick's furious endeavors with a double-bitted axe was the let-down when a job was over. Then days on end he would relax, lying, like as not, if the time was a soft, mellow autumn, stretched out on the ground in a grove of butternut trees and languidly watching the squirrels as they skun up a tree to loosen a cluster of nuts, only to dash down and try to beat Dick, now suddenly all energy, to the loot. Dick could beat the squirrels by yards, but the little fellows never stopped trying.

The tall, ungainly Dick was never a canal man, but he was so constantly trapsing up and down the upper levels that he contributed richly to the lore of the boatmen. It was on a down-trip from Boonville, where he had seen his first automobile, that, telling a lock-tender about the experience, he went out on a limb with a prophesy, "I tell yuh, Bill, it won't be more'n a few years till dead people 'll be ridin' around in them things." With such enthusiasm was the prediction received by the lock-tender that Dick forthwith went about telling one and all of what they might look for, and when one day a motor-driven hearse actually appeared in Boonville he declared that they had got the idea from him, a view in which one and all, to bolster his morale, did much to encourage him. It is the author's happy privilege to record that Dick took the business serenely; his disposition did not curdle, nor did he lose such faith as he had in human nature. It is a privilege also to say that he was susceptible to the finer aspects of human contacts. When occasion arose Dick could throw high-falutin' words around with the best of them. As when, one afternoon, sitting by the stove in the back of Pete Gill's grocery in Boonville, a kid came in and asked where Karl was. Karl was Pete's delivery boy, and at that moment, waiting for his next call, was in the back room doing his home work in arithmetic. The situation called for something scholastic and Dick, pointing his thumb, came up with "He's in there reckonin' his sums!"

11

Trials of Traffic

To own a boat and have to depend upon it for a living required of a man a hopeful attitude toward life and an unending fund of patience. Or both. The season of navigation lasted through but seven months of the year—that is, from May 1st to December 1st. An early spring might bring the opening date back to the middle of April and a late autumn hold the canal open until ice in the canal brought an end to operations. For on the Black River canal the day of actual closing seldom jibed with the official closing. Homing boats bound for Forestport, say, might be climbing the canal from as far down as Westernville or Rome on the closing date, and indulgent officials always permitted them to get to their winter parking places.

In the canal's later years the opening and closing dates were set for a few years for June 1st and October 1st, for reasons of economy. This arrangement was dictated by the policy by which the State's attitude toward the Black River canal, as the result of pinch-penny economies in the maintenance department, had reduced the canal to an all but unnavigable condition, and around the beginning of the present century the Canal Board was surprised to discover that boats were fast disappearing from the canal. One year the Board was deploring the dropping tonnage figures and considering enlarging the canal to remedy the situation, and the next, to offset the conditions which its policies had created, cut short the season by two months.

In any case never could the boatman look forward to uninterrupted navigation, unless he was fortunate enough to dodge breaks and waterless levels in the dry spells by being down on the Erie bound for New York when such catastrophes occurred. And if it wasn't breaks and lack of water it could be one or another of several other kinds of accidents.

In 1873 navigation in the Port Leyden section was suspended for several hours when a house fell into the canal. The record does not say whose or what kind of house it was. Gates had a way too frequently of going out, as when in 1872 lock 4 in Rome lost a gate and navigation was suspended for two days, and the following year lock 32 up in the Lansing Kill valley. Boat sinkings claimed their toll; sand-laden boats, in the later era of heavy sand haulage, were frequent casualties, a fact due not so much to the heavy loads as to the fact that too many aged and aging boats were in the business. In one week in 1912 two sand boats went down, the "L. W. Fiske," between locks 63 and 64, the Marsh brothers captains, and the "John House," Will Golden captain, between locks 47 and 48. The owners were advised to abandon the "Fiske" as being too old for sand. It was explained that the great weight and solidity of the cargo, and the resulting momentum of the boats, caused their hulls to crush in easily when they collided with any obstruction forcibly, especially when the boat was weakened by age.

The same reason cannot be given for the frequent sinkings of boats carrying pulp wood, and yet, perhaps because so many were in the traffic, accidents were frequent. The "Joseph Ano," a Forestport boat, Fred Race captain, who had bought the boat from Roscoe Clark, got jammed in the feeder; how and why is not recorded, but there the "Ano" was, crosswise of the canal. In attempts to right the boat it sank, causing a traffic delay of nearly twenty-four hours. When bridges fell into the canal, and they sometimes did, almost anything could happen. In 1875, near Westernville, a farmer's wife was "crossing a bridge driving a team hitched to a mowing machine," as the report relates. "Mrs. V. had three ribs broken, but was otherwise uninjured"! (Exclamation point is the author's.) "The team and machine were gotten out of the canal uninjured."

Accidents involving locks and lock equipment were perhaps more frequent than from any other single cause. As early as 1864 Daniel C. Jenne, in the report of the State Engineer and Surveyor, stated that "obstructions to navigation have not been of frequent occurrence except occasionally by the breaking of a lock gate. More hindrances occur to navigation from gates and valves than from any other source. They are so badly worn that it is impossible to put them in good order," and Jenne urged that new ones be installed in practically every lock south of Boonville. Traffic delays incident to faulty locks might be for a day or two.

In the later years of the canal, repairs to gates had a high priority rating each winter and spring in preparations for the following season. In the reports of Fred Woolley, Superintendent from 1915 to 1918, the replacement of worn and defective gates had a prominent place, as when in 1918 locks 3, 4, 14, 18, 19, 24, 25, 30, 31 and 33 were fitted with new gates. For that matter, it was in the later years, when it had become clear that the canal's days were numbered, that Superintendents were most assiduous in making repairs directed towards the *prevention* of trouble. Thus in 1902 Bert Calen not only put in a goodly number of new gates, but fifteen mitre-sills and seventeen balance beams; foundations of locks began to be strengthened in a major way as part of the preparation for each season of navigation—even neglected lockhouses began to blossom with coats of paint, gaping windows to get new panes, and the canal banks to be cleared of brush and weeds.

It was not always worn lock fittings: boats inside a lock took their full toll. Long ago, in the summer of 1854, the "Northern Light" had one of its bad days when, in getting through lock 32, it left the gates a wreck, and as if this were not enough it went into lock 31 and performed the same feat, with a consequent three-day delay in navigation.

Low water got in its licks too in unusually dry summers. The Black River storage system as completed had tapped every lake and pond of consequence in the region of the Black River head waters, including White Lake near White Lake Corners, the present Woodgate. The supply of water was adequate for normal seasons. Moreover, in 1881 the Fulton Chain of Lakes was dammed, the Moose River carrying the water to the canalized river at Lyons Falls, while the Beaver River reservoir was completed in 1887, going a long way towards keeping traffic moving on the lower river and the wheels of industry turning in Watertown, Carthage and the villages in between.

The last of the Black River reservoirs to be built was the upper Forestport pond, completed in 1893. The contractor for this job was Forestport's own Phil McGuire, the greatest shipper of pile and spar timbers. Phil ("Black Phil" to one and all, to distinguish him from his cousin, "Red Phil" McGuire), had come to this country from Ireland as a boy, and who, without benefit of any education whatsoever or a bank account to start with, had accumulated a fortune, and the ungrudged respect of the entire countryside. Not only was he in the big-time lumber business, but he ran a farm of three hundred acres, raising immense quantities of potatoes for canal shipment, and owned grist and woodpulp mills, a cheese factory and a tannery, among his many and wide variety of interests.

Phil at that came near not getting the reservoir made. The story is of the gang of Italian laborers brought in to work on the project. Arriving at the scene of operations, down back of Alder Creek, from the vantage point of the high bank they looked down on the wide, fast-flowing stream, and their leader, deciding that the hole was too big ever to be filled up, quit with his men and went back to Utica. Of quite a different nature was the huge Russian character who got a job on the Delta dam, only to be fired because when lugging cement to the mixers he insisted on carrying three bags at a time, and so kept the foreman's schedule constantly out of kilter.

The dam, it may be stated, was fortunate in getting started in an atmosphere of comparative calm in Forestport, such as did not always prevail. A year before the Italians left in a huff the Forestport correspondent of the Boonville *Herald* in one of her weekly pieces wrote that "our happy friend, Jack McElroy, of the Albany Medical School, has been for the past week enjoying the vicissitudes of his native clime." This could have referred to any one of a number of things; it may have alluded to an incident referred to the previous week by the Hawkinsville correspondent, who had it that "the umbrella mender that pervaded our streets Saturday, and who was reported as having met a violent death in our midst, was last seen on the tow-path headed for Forestport." If it can be proved that he never reached Forestport that lets him out.

The cause of the vicissitudes that afflicted the village could also have grown out of the Dutch Hill War. This affair got going when Joe Kreyer, again according to the *Herald's* Forestport correspondent, who lived over across the canal on Dutch Hill, and whose convenient pump was resorted to among others by boating people at such times as it was inconvenient to visit Mulchi's taproom over town, started to charge a small tax for its use. Immediately the community was up in arms and declared war, according to the correspondent. It made no difference that Joe explained that the levy was imposed for the laudable purpose of keeping the pump in unfailing repair. A few of the citizenry took his side, declaring that users of the apparatus should come up cheerfully with the fee, but little good it did Joe. Finally Lon Denton, thinking to compose the emotions of the populace, said something about to fee or not to fee. That did it, but in an unexpected way. It had the effect of taking the heat off Joe and pitting both sides against Lon. With both sides united in this common interest the ruckus was not long in subsiding.

Even with the new dam, however, there was not enough water in exceptionally dry summers. In 1895, as a matter of fact, two years after the dam was completed, the canal north of Boonville had to be closed. Every reservoir gate was thrown open, and still the feeder level was twenty inches under normal. In Forestport twenty-five boats were tied up, and in Boonville were grounded the "Scudder Todd," "Watson M. Shaw," "Charles P. Evans" and "George Homan," all loaded with brick for the paper and pulp mill being put up at Lyons Falls, the "Eddy and Charley," with a cargo of grain, and six others loaded with lumber.

And so it was up and down the canal. Old boatmen recalled other dry years, notably 1873, when eighteen boats were laid up in Boonville, and as many more at Forestport. One captain who somehow managed to reach the summit level from Lyons Falls, was four days on the way.

A boatman laid up in the vicinity of his home at such times did not count the delay a total loss, for he could pitch in and harvest his hay, or whatever crop might be ready for taking in at the time. Or if he was held up near somebody else's farm that needed a hand he like as not would come to the rescue.

Canal-filling slides were not too frequent, but when they came the boaters always took a beating. Slides were confined pretty much to the Lansing Kill valley, whose steep shale slopes, loosened by heavy rains, might send thousands of tons of debris into the canal, filling it for rods at a stretch. The most devastating occurred one season immediately up-canal from the Five Combines, two weeks being required by State Shop crews from Boonville and North Western to put the level back into operation. The hillside at that point was given new contours, such was the volume of the slide, that are evident to this day.

It was breaks, however, that most constantly harrassed navigators. A break might be caused by any one or a number of conditions—by heavy rains that weakened canal banks; by sink holes, into which a section of a bank, and even of the bottom of the canal itself, would completely disappear; by seepage into and through a bank until what soon became a trickle presently was a rush of water that carried sometimes as much as three hundred feet of the bank with it.

During its first years the canal was fairly free of breaks, but by 1855 they began to show up with uncomfortable frequency. In the July of that year, in a break below the Sugar River, at a point opposite Hurlbut's mill, where the Boonville power plant now stands, one hundred feet of the tow-path were carried away to a depth of eight feet below the bottom of the canal. Sand, flood-wood and other debris were carried into the mill, causing vast damage to machinery and weakening the foundations of the structure. It was two weeks before traffic could be resumed. At practically the same hour a minor break occurred just north of the aqueduct across the Sugar, near the four combines. Repairs were made, however, before the bank at Hurlbut's mill was repaired, so that it caused no delay in traffic.

The level between locks 31 and 32 was another trouble spot. The high slopes, rising directly from the canal, would bring, after a freshet, a flood of water that, after filling the canal to overflowing, broke through the tow-path bank and carried great sections of the level into the Lansing Kill twenty-five feet below. Engineers declared this to be the canal's No. 1 trouble spot.

12

The Breaks

An unusual accident occurred in Boonville on September 22, 1865, when the waste-weir gave way, with some astonishing results. The bridge over the weir, weakened by the wreck, fell while Dan Hulser was crossing it with his team, killing both horses; the entire canal was emptied and traffic halted for two days; the uneven canal bottom caused boats to open at the seams and was responsible for a hole to be poked through the bottom of a new boat, the "Bill Camel," and weakened a number of others; a load of salt, consigned to S. E. Snow, commission merchant, had just reached the basin when the break occurred and when water was let into canal it also made its way into the cargo and left the load a total loss.

A wave of indignation swept through the village when it was learned that the timbers comprising the weir gates had all but rotted away. Blame was put where it probably belonged, on the contract system of making canal repairs. Under this system annual contracts for making canal repairs were let out in competitive bidding. Opponents of the system, particularly the canal engineers, had from the start opposed it. Year after year State Engineer and Surveyors had attacked the system as fostering conditions that made breaks and deteriorated canal equipment inevitable. The State Canal Commissioners, however, persisted in their mistaken notions of economy, until a mounting roll of disasters and near disasters compelled the abandonment of the system.

Readers will be gratified to know that a purse was made up in the village whereby Captain Hulser was able to buy a new team of grays.

During the night of June 20, 1901, Hawkinsville escaped a major disaster by minutes. Excessive rains during the night had brim-filled the feeder, and around three in the morning the tow-path bank gave way near Second Street. The roar of water pouring through the thirty-foot gap opened up awakened Caleb Carr, who lived within yards of the break. Caleb went into quick action and found George Cannon, the bank watch; George opened the gates of the waste-weir and telephoned Forestport for the gates there to be closed. This prompt action saved the village from almost complete destruction, though it was not in time to prevent deep gullies being torn in First and Second streets and across yards, and the demolishing of fences and small buildings in the water's mad rush to the river. No fatalities occurred, and such was the fervor that Superintendent Bert Calen and his repair crew put into their efforts that water was let into the feeder thirty-six hours later.

The most savage of all Black River canal breaks, and one of the worst in the history of the State's canals, occurred north of Forestport on July 23, 1897, when four hundred feet of the tow-path bank and fifty feet of the feeder bottom went out. The disaster occurred at a point where the feeder is carried upon the side of a bank, seventy steep feet above the river bottoms. Practically the entire engineering staff of the Middle Division of the State Department of Public Works, to which the Black River canal was attached as an administrative unit, came to the scene and went into action. After thirty days of almost superhuman efforts navigation was resumed. The job was organized and carried out in so superatively effective a way as to deserve here an excerpt from an interview with Thomas Wheeler, of Utica, Superintendent of the Middle Division, which appeared in the Utica *Herald:*

"At first seventy-five men were placed at work, and this number was gradually increased to 1,700. Gangs of men were kept at work night and day and the large hole in the ground was speedily filled up. Just how much dirt was used in filling in the break I could not say, but a conservative estimate would be about one hundred thousand yards. The first work done was the building of a sheet piling; two rows 53 feet high were constructed across the cut. A half-arch core was then constructed, extending from one end of the break to the other. Sand and gravel were used in filling the core, and these were puddled, which consisted of mixing and pounding, which makes the earth more solid.

"The sand was secured near the break by an endless chain of teams. No team stopped to be loaded, as there were so many men at work that as the wagons passed by each one threw in a couple of shovelfuls of sand

and when the wagons reached the end of the line they were loaded. The sand was secured from two sandbanks, the gravel from a pit two miles from the break. There were 250 teams at work, and each team drew six loads of gravel a day."

For their part in all this the workmen on the job were paid at the rate of $1.65 a day, the teams at thirty-five cents an hour, Sundays time and a half. The cost of the operation was $62,781.78.

Then came the spring of 1898, when on May 23 came a second break, surprisingly close to the scene of the one the year before. This one suspended navigation for twenty-one days and entailed a repair cost of $50,764.47.

This time the engineers did more than make repairs—they became acutely aware of strange similarities in several respects between the two breaks. By the time they had readied the feeder once more they held strong suspicions that both were man-made, "inspired by persons who would be benefited," as one of the engineers put it, "by having larger sums spent in the locality."

Came the summer of 1899 and an under-cover investigation got under way, but scarcely before a third break came, on September 18th, at practically the same place as the preceding one. Seventeen days, with crews working night and day, were required to repair the breach, at a cost of $17,089.72.

State Superintendent Partridge now went into action. He brought in operatives of the Pinkerton detective agency, who worked with such enthusiasm that nearly a score of men were arrested, several of whom, scenting the brewing of trouble, had decamped, one being caught in scenes as far distant as Michigan. Indictments were returned against thirteen, of whom one, proprietor of a hotel, first receiving a sentence of not guilty was tried a second time, convicted, and given four years. Another drew a prison term of three years; three entered pleas of guilty and drew terms of one year each; three others, upon pleas of guilty, received heavy fines; three others were freed, the State having no evidence against them except statements of those who had confessed, and the remaining two, having turned State's evidence, were released on their own recognizance. And so came to an end the most sensational episode, or series of episodes, in the history of the canal.

A disaster of a different kind descended upon the canal in April,

1869, when the entire Black River valley was devastated by torrential rains. At North Lake the dam went out, releasing a head of more than four hundred acres of water, thirty-three feet in depth, that poured down through the valley of the North Branch of the Black River, striking at Forestport with a fury that left the river-front properties a shambles. The State dam, and all of the bridges in the vicinity went out, a number of saw-mills and other manufacturing establishments were wrecked, and, booms in the pond giving way, thousands of feet of logs waiting to be made into lumber, were carried down-stream. Mills at Hawkinsville suffered an estimated loss of $50,000; White's bridge in Leyden was carried away; at Port Leyden the great Snyder tannery and the Port Leyden Iron Company reported costly damage, bridges were destroyed and one life was lost. And so ran the toll of destruction the entire length of the river.

Public indignation rose to fever heat when it was learned that the disaster followed the replacement of an experienced dam-tender at North Lake by a new hand, a novice in the business, the change, it was charged, having been made in the payment of a political debt. Not sensing the meaning of the speed with which the reservoir was filling under the impact of the heavy rains, a notable characteristic at North Lake, the new man took no steps whatever to avert disaster by opening the gates.

Water in the river had barely started to go down when property owners, from North Lake to Watertown, were busy filing damage claims against the State, totalling, when all were in, an aggregate of $700,000, with 2,370 claimants. The figure, it would seem, was modest enough, even making allowances for chicanery resorted to by such claimants as seized the opportunity to pick up a little easy money. The Canal Appraisers, however, were nothing if not diligent. In their exertions they were tireless, sifting evidence in each case, but never for a moment forgetting that they were the servants of the State. A goodly proportion of the claims were rejected, and the others were granted in amounts running from nominal to sub-nominal. Dissatisfied claimants appealed their cases to the Canal Board, but little good it did them, the Board in practically every case upholding the decision of the Appraisers.

A new dam had to be built at Forestport and the canal, which had suffered from the freshets, particularly in the Lansing Kill valley, where levels and culverts were choked with sand and other debris, must be conditioned, and it was June 1st before the canal was opened to navigation.

Boatman, grounded by breaks on the bottom of the canal, took their enforced leisure with a good deal of philosophy. Roscoe Clark remembers a veteran of the canal who, while repairs were being made on the Forestport break of 1897, reckoned as how you might as well have one big wait a month long as a lot of short ones. One old-timer said heck, he wasn't going nowheres no how but to New York, while another, with a show of satisfaction, declared that now he would have a month before he would have to negotiate the cussed "Sixteens." The Sixteens was the series of locks of that number that lifted a boat out of the Hudson at West Troy and set it down on the high land above, headed for Schenectady. They were manned by the roughest, toughest bunch of lock-tenders found probably in the State's canal system. Approaching the locks the boatman nailed down everything in sight. Even then things disappeared. If the cargo was of lumber the boatman was lucky if, while his back was turned for the most fleeting of instants, a few of the choicest boards did not, by some sleight of hand, disappear. Or tools, if the captain had not got them out of the way. "Scale money," canal language for gratuities, which in turn is uppity language for tips, had it ever been offered, would have done the Black River men no good; quiet-going men, not obstreperous like their counterparts from some other regions, according to the Black River boatmen themselves, they were the special object of the attentions of the men of the Sixteens. The boatman's only defense was eternal vigilance on the way through, unless he was Peter or Allen Phillips, whose strong-man reputation secured for them an unchallenged passage of the locks.

Traffic delays also promoted the education of new drivers in the lore of the canal. Hank Dakin, from down the river, was laid up one time near lock 32 and says it was there that he first heard from the captain about a pair of female brigands, known thereabouts as Jane and Susie, who, arriving from the shooting West, had proceeded to terrorize the North Western-Dunn Brook sector. The girls came to grief when one dark night, garbed in men's clothes and each astride a charger, they commanded Hugh Dorrity, at the point of guns, to put up his dukes. He obliged, and they then demanded his roll. He declared he had no money on his person, and told them who he was and said he recognized their voices. They believed him on neither point and allowed him to strike a match in his face to make good his own identity. By the same strategem he got a good look at the countenance (such as it was, from all of the accounts) of each of the robbers. This was enough; the desperados let Hugh proceed, after cautioning him to keep it all under his hat. Next day Hugh got the law on them; the sheriff, after looking everywhere else, found them and their horses hidden in an abandoned canal boat. They put on a vicious battle, but the law was finally victorious, only, however, after much clawing and screaming by the ladies and some of the sheriff's fancy cussing. It is pleasant to know that a stretch in jail helped the culprits to fix upon less violent ways of life.

Erlo Capron adds for the author a footnote to Hank's story. Susan, when she had reached an age when even desperados seek more moderate modes of conduct, was the object of occasional torment by youthful pranksters. On the occasion of their last endeavor she dashed out onto her porch flashing a pair of six-guns, skilfully putting a bullet through the hat of one, barely missing the feet and other portions of the anatomy of others, and all in all displaying nicely the precisions of her old-time techniques. The aggregation fled in terror and found refuge in the nearest asylum, which was the local bar. They were still catching their collective breath when Sue suddenly appeared in their midst, brandishing the revolvers and again placing shots with uncanny accuracy, to the accompaniment of the quickest exits, the bar-keep afterward said, that the place ever had seen. Thereafter she was permitted to live in a degree of peace that must have been hard for her to endure.

Hank also had from his skipper the sad tale of "the meanest skunk that ever worked on a scow." Jack Allen the name. Jack shipped downriver for Anthony Rape, and on his second season out annexed a pair of the boss's boots. Anthony had him arrested, which naturally made him as mad as two hornets. In jail he didn't feel much better about it when told that he would be tried before the jury. He asked heck, couldn't he be tried after the jury so's he could see how they did it. The sentence of three months given him by the judge did little to mollify him.

Hank recalls Ezra Merry, when they were jawing one day at the Bush's Landing lock, telling him about Jim Ervin, captain of the steamer "T. B. Basselin," one time down off Lowville losing a pocket-book overboard in some fifteen feet of water. Recovery, by any means short of a shark coming up and handing him the purse, being out of the question, according to Ezra as passed on through Hank, Jim sent to Ogdensburg for a diver to come with one of them diving dinguses. Well, sir, he went down easy like, and when he come up by jiggers if he didn't have it. Not a ten-spot was shrunk, like it is today. There was three hundred and seventeen honest-to-god dollars, and when Jim sees the bill he has to pay he says, "Hell, it'd been cheaper to go up to Ogdensburg and drop it in the S'Lawrence."

13

Timber Becomes Big Business

The Black River Canal until the 1890's was essentially a carrier of timber, lumber, and other timber products, such as wood-pulp and tan bark. Prior to the extension of the Utica and Black River Railroad north from Boonville, the canal had the freight hauling business to itself. The railroad when it went on through took a cut of the general tonnage, but the special field of forest products still depended upon the canal. One reason, and an important one, was geographical. The forests were on the east side of the Black River, and timber and lumber could be loaded from the river bank directly onto the snubbed-up boats. From Lyons Falls the canal boats even went up the Moose River to take on loads from the great Gould mill a mile above. In a similar manner at Naumberg boats were hauled up the Beaver River to take on lumber from Theodore B. Basselin's Beaver Falls mill—here the mill met the boats part way, a tramway carrying the lumber down to the uppermost point of navigation.

The canal, in its first full year of operation, 1851, reported clearances of 25,320 tons; in 1861, ten years later, this had grown to 69,930 tons; in 1871 to 89,560 tons, in 1881 to 100,233 tons, and in 1891 to 122,111 tons. These decades cover the great timber and lumber years. As a matter of fact, beginning with 1892, with 115,469 tons, the figures, with the exception of a half dozen years when heavy traffic in sand and stone ran the tonnage up, progressively declined until the canal's abandonment in 1922.

Of the annual tonnage during this period of 1851 to 1891, lumber and timber account for an average over the years of better than eighty per cent—in 1857 timber products in the form of lumber, shingles, staves and timbers reached an all time high, with 93 per cent of the total tonnage of 69,135 tons. The low percentage of non-timber products during all these years cannot be ascribed to the coming into the North Country of the railroad. The road had reached Boonville in 1855, terminating here until 1867, when, now as the Utica and Black River, it moved on north to Lowville in 1868, and Carthage in 1871. It never, however, competed with the canal in the hauling of forest products; its important contribution to the region was a stimulation of commercial activity and production in agricultural and industrial goods. In these fields the tonnage went for the far greater part to the railroad, although in the shipment of potatoes for the New York City markets, and, in its later years, sand and rock, the canal took a very considerable cut.

These facts, in any discussion of the decline in traffic that led to the canal's ultimate abandonment, make it clear that the railroad had no appreciable effect upon the lumber and timber tonnage of the canal. The railroad did cut into canal tonnage in other products—in fields which the railroad was building up; the canal did get a slice of this new tonnage, but it could not get enough of it for the reason chiefly that the movement of boats was slow, and that in the later years the delays of boats occasioned by the bad physical condition of the canal, a condition that became chronic, forced shippers to turn to the railroad.

The original sponsors of the canal had seen the canal chiefly as a carrier of forest products, and so long as the timber lasted the canal more than made good their claims; with the disappearance of the forest little remained in other fields that the canal could turn to except potatoes and sand. Potato production, however, like the forest, declined, while sand alone could not maintain a canal. After 1892 the canal, so far as any foreseeable future was concerned, was through. It was not the railroad, however, that brought about this regrettable fact: the canal simply had fulfilled its destiny.

The production and shipping of timber and timber products, however, throughout half a century makes a spectacular story. During that period the vast timber lands covering the entire region of the Black River head waters beyond Forestport, and lying east of the Black River all the way from Forestport to the Beaver River and beyond, disappeared. Where small, isolated mills in 1851 were operating between Lyons Falls and Naumberg, huge gang saw mills were to come in, steam would be installed where water power was inadequate and uncertain;

mills that had made lumber purely for local use would give way to enlarged operations that sent millions of feet each year to Albany and New York markets.

Two years before the arrival of the canal Henry and Marshall Shedd built a saw mill at the lower falls of the Moose River, a mile above Lyons Falls, later adding wood-pulp equipment. In 1866 the plant was sold to Emory Allen, for whom a canal boat had been named, and Newton Northam. This plant operated through a period that saw several changes of ownership, the mill finally passing into the hands of G. H. P. Gould, who built the property into one of the greatest mills in the North Country. At Lyonsdale C. L. J. Ager conducted a saw mill enterprise alongside the Ager and Lane paper manufacturing business which had been established in 1848. Above Lyonsdale other mills were being operated by James Hyland, of Boonville, and Henry Brown.

In the town of Greig Richard Carter in 1853 put up a gang saw mill on Otter Creek, and a number of smaller mills soon were in operation in the town. At Beaver Falls the Prince mill, also with gang saws, was adding its output to the stream that by the 1880's was to become a veritable flood of lumber. And on Sugar River, at Port Leyden—in a score and more of places, as a matter of fact, other mills, of smaller or greater capacity were got under way at the beginning of canal operations.

Forestport in 1851 could boast of two saw mills, one of them in spasmodic operation. Already, however, ambitions began stirring dreams among the younger men who were coming along. Young Henry Nichols, for one, who was learning the turner's trade, and who, soon after marrying and buying a farm of three hundred acres, set up a saw mill— in a small way at first, but the beginning of a business that was to grow and make him, in the flood-time of lumber, one of the big operators. Along about the same time Phil Hovey was fixing to marry Julietta Kilmer, which he did in 1858. He too was soon in lumber, modestly, but in a way that was to make him one of the greats, requiring for delivery of his wares a number of canal boats which he built to carry his lumber to market. In that early time, too, two lads in different parts of the State, Alonzo Denton and Nathaniel Waterbury their names, were, although unaware of the fact, being conditioned for service in the Civil War. At the close of the struggle between the States turns of events brought them to Forestport, where they became partners in a general store; to Lon they brought a wife in the person of Carrie A. Waterbury, Nat's sister. The combined energies of the two young men could not be confined to a store, and presently they were in lumber, the firm by the time the 'Eighties were reached having become one of the greatest producers of lumber in the region.

Philip McGuire's advent in Forestport preceded that of Denton and Waterbury by some two or three years; he arrived after a year spent in the woods above Lyons Falls. He was, as near as these researches have been able to discover, twenty-six years of age. Three years later he was at work upon an idea that was to become the chief of a wide variety of McGuire activities—the production and shipment of spars and piles for the city market—long (as long and straight as could be found) timbers of spruce adapted for use in making spars for sailing ships and piles for building the miles of piers along New York's water front.

These timbers, assembled by expert cribbers into "cribs," a kind of raft made up with three layers of timbers, contained as many as sixteen timbers, the number varying with the size of the pieces. Five or six cribs were assembled in a "tow" to be hauled by the tower's team, which, according to all accounts, liked the business as little as the lock tenders, and little less than the captains of boats which might be lining up below a lock impatient to get through. For a tow had to be locked through crib by crib, a monotonous arrangement at best. Moreover, owing to their great length (for a crib might completely fill a lock) cribs were less manageable than boats and were a constant hazard to lock walls and lower gates.

14

Down-River Lumber

The big era in lumber production may be regarded as beginning in 1881, when for the first time Black River annual tonnage went into six figures. This was a jump to 100,233 tons from 75,308 in 1880. Once previously six figures had been almost reached—in 1870, with 96,329 tons; promptly, however, tonnage went into a relapse, in 1877 reaching low for that decade, with 63,286 tons.

From 1881 to 1893 six-figure tonnage was maintained, the latter year reaching the all-time high for the lumber era with 143,561 tons. The volume and the year had no relation to tonnage for the New York State canals as a whole during the same period: the peak year for all canals as a unit was 1872, which turned in a total of 6,673,370 tons.

If 1893, with 143,561 tons, was the Black River's climax, the next year yielded an astonishing anticlimax, for 1894 turned in a scant 56,024 tons, a dramatic announcement that an end had all but come to Black River timber. Two or three years thereafter would turn in high tonnage reports, as 1910, which reported 175,966 tons; in that year, however, the Delta dam was under construction and the canal was lined with boats carrying stone and sand for its construction. It was sand, as a matter of fact, that during the later years kept the canal going at all, sand with an assist from firewood. But that is a story belonging to another and a later discussion. To this present chapter belongs the down-river part of the lumber story.

In 1879 Lewis County shippers and boatmen held a notable conclave, a meeting held to thrash out the whole business of toll rates: the effect of toll schedules upon the volume of canal business, and means by which relief could be obtained from the State. Theodore B. Basselin, of Croghan, was chairman of the meeting and G. H. P. Gould, of Lyons

DOWN-RIVER LUMBER   101

Falls, the secretary. T. J. Lewis, of Beaver Falls, Hamilton Wilcox, of Lowville, and Gould were selected as a committee to draft resolutions.

The toll situation was not a simple one; a law of 1875 provided that in no year could a canal's administrative and maintenance costs exceed double the amount of tolls collected. The Black River canal was chronically in a bad state of repair, and the problem was to get tolls down and at the same time keep maintenance up. The meeting did the best it could: it adopted a resolution petitioning the Canal Board for relief and appointed Samuel Garmon, of Watson, a Committee of one to work with Senator Charles A. Chickering, of Copenhagen, in gathering data upon which new rates could be established. The Senator was cooperative to the extent of introducing in the Senate a resolution calling for a reduction of tolls on butter and cheese, a commodity that was going to the Utica & Black River Rail Road anyhow, and on slab and cord wood, which figured little in down-river commerce.

For the purposes of this chapter the most important feature of the meeting was the unveiling of a table giving figures for the estimated production in lumber for the then current year—and it was toll rates on lumber with which the meeting was particularly concerned. The detailed estimates follow (all figures are for board feet):

Frank B. Ward, 1,000,000; Van Amber Brothers, 3,000,000; T. B. Basselin, 5,000,000; C. J. Farney, 200,000; Michael Andrews, 200,000; Peter Yancey, 200,000; A. M. Searles, 300,000; A. J. Pasenger, 500,000; Peter Beller, 400,000; William Crumb, 700,000; Jacob Petrie, 1,500,000; Crawford and Company, 500,000; Ralph Beach, 500,000; Peter Beller, 400,000; Frank Sperry, 500,000; William Glenn, 1,500,-000; George H. Crandall, 600,000; Duane Norton, 1,500,000; T. J. Lewis, 5,000,000; Wessel Gallup, 1,200,000; Charles Thorpe, 1,000,000; Youngs and Wilson, 300,000; Royal Bancroft, 500,000; Wilson Higby, 500,000; George Van Aernam, 200,000; Rockwell Abbey, 200,000; Joseph Northrup, 500,000; Hiram Peak, 500,000; Carthage parties, 200,000.

In this sector Basselin (pronounced thereabouts as Bazzelin) was arousing wide interest by reason of his flair for the dramatic and a kind of Napoleonic touch with which he conducted his operations. In 1885 he built the mill which identified him most closely with the canal. The structure was put up on the west bank of the Black River, part way between the mouth of the Beaver River and the Castorland bridge. He already had two mills, at Beaver Falls and Belfort. He resided in Croghan and could keep an eye (and it was a sharp, shrewd eye) on his varied interests, which included tanning and wood-pulp production along with lumbering. He owned, as has been pointed out in a previous chapter, a fleet of canal boats for getting his goods to market, and steamers to tow them to Lyons Falls. He never had enough canal boats, however, and relied upon other boats in no small degree.

An extra-curricular activity of Basselin's was politics. Pertinent to the history of the canal is the fact that, a Democrat, he associated himself closely with two men who played important parts in the administration of the canal—Samuel Garmon, of Watson, and James Galvin, of Carthage. The advancement of Galvin to the superintendency of Section 2, which at the time ran from Carthage to Boonville, raised Republican eyebrows and stirred murmurings. It was, they claimed, a typical triple-play: in this case Basselin to Garmon to Galvin. Galvin was appointed for the year 1883 and continued until 1887. His administration was of an exceptionally high character; his team mate through all these years on Section 1 was Peter Phillips, the pair of them, working as a team, giving a high quality of service that the canal too seldom enjoyed. Later on Garmon himself became Superintendent, and again suggestions of political shenanigans were heard. Few canal appointments, however, but were politically made; no attempts were made to keep the fact hush-hush, and consequently everybody accepted the fact without any undue show of cynicism, and most canal men with a good deal of humor. Whenever the political complexion of the State administration changed, lock-tenders and scow and State-shop workers began to pack their belongings and consider plans that would carry them through until another overturn in the political character of the administration might return them to their jobs. One old lock-tender, who by some miracle had survived two such overturns, on being apprised that the election just held might mean separation from his job, replied, "I'm not worryin'; I could take my valise and go out right now." And that was the way it was.

It can be said in all truth that the office of Superintendent was, with surprisingly few exceptions, filled with able men; if in a given year the appointee must be a Republican or a Democrat, every effort seems to have been made to select a man who measured up to the job.

It is a fact, worthy of note in passing, that Basselin, Galvin, Garmon and G. H. P. Gould, the lumber genius of Lyons Falls, were the Black River's contribution to the membership of the State's first conservation commission.

The same year that saw the Basselin mill rise on the west bank of the Black River witnessed the entry into the field of a newcomer, who built his mill just back off Independence Creek, with a tramway to carry his lumber to the loading bank on Black River, three-quarters of a mile below. This was Henry Abbey. Henry had cut his lumber teeth in a mill owned by his father, Rockwell Abbey, at the southern edge of the village of Greig. After the death of his father Henry went over to Partridgeville and leased the old Partridge mill, buying his timber from Lyon and Gould, a firm conducted by G. H. P. Gould and the heirs of Lyman R. Lyon: Mrs. Florence L. Merriam, Mrs. Mary L. Fisher, and Mrs. Julia L. DeCamp. The arrangement was a successful one, and in 1885 Henry built the Independence Creek plant and sawed lumber for the Gould-Lyon interests on a per-thousand-feet basis. The annual output was to run as high as six and seven million feet a year. Later, in 1895, Henry was appointed Superintendent of Section 2, and did so well that he was continued through 1896 and 1897.

Another year and the 1879 list would not have included George H. Crandall in its list of lumber operators. For in 1880 Crandall sold his mill, situated at what had come to be known as Crandallville, on Independence Creek, to New York's greatest firm of lumber dealers, Dannat and Pell. The *New York Merchantile Review* at the time described the company's New York yards as being "at the foot of Broome Street, East River, covering in all forty-two city lots. There is a water frontage of three hundred feet, at which many vessels can lie at one time to load and unload. These yards have a capacity for the storage of twenty million feet of lumber. They handle probably 40,000,000 feet per annum, and year by year the total becomes greater."

In Lewis County, the article goes on to say, "Messrs. Dannat and Pell have a tract of land of over 2,500 acres, and are now negotiating for the purchase of 1,000 acres more. They have established a village there, the inhabitants of which are men employed by the firm in cutting and dressing the lumber and their families" a bit of syntax, that last phrase, which the reader can work out for himself. "About a year ago the inhabitants called a meeting and after some discussion the name Dannatburg was adopted. There is a church now being built and a club is being organized for social purposes."

Crandall, it may be noted, gave his name to a canal boat, the "George H. Crandall," owned and run by Chauncey Wetmore, of the Watson Wetmores, and afterward owned by Jerome Salmon. Chauncey while running the "Crandall" lost his life in an accident at Sprakers, down on the Erie, that saddened the canal fraternity as few accidents had done. But twenty-three at the time, he was immensely popular, as was his young wife, daughter of George F. Beach. About to step down into the cabin of the "Crandall" she lost her footing and fell into the canal; the young man plunged in to save her, but was dragged under as she clung to him, and both were drowned.

The Ralph Beach of the 1879 list was cousin of Nelson J. Beach, who has already appeared in these pages, and father of Andrew J. Beach, who built a series of boats under the Beach aegis. Ralph himself built one boat of which record remains, the "R. Beach Jr. and Sons."

In including Crawford and Company the list did not tell the entire story of the firm's activities. A striking product of the organization was a tanning extract made from the bark of the hemlock. From the four thousand cords of bark which were used each year 3,500 barrels of the extract were made, to be shipped to leather manufactures chiefly in Connecticut and Massachusetts. The advantages of the product were, convenience in shipping as compared with bark, and also a saving in shipping costs.

When the bark was removed the Crawford saws cut the hemlock logs into lumber. In a way the lumber end of the business could be considered a by-product of extract production manufacture, in the sense at least that the plant, established on the Chase Lake outlet in 1871, was built primarily for the production of the extract. Prior to the coming of the canal hemlock had been cut, de-barked, and then allowed to rot away, in the absence of means of getting lumber to market. The canal had changed all that, and so long as the timber lasted a tannery site would be found not far from hemlock lumbering operations. In the Crawford instance one establishment served to process both the bark and the timber from the firm's 7,500 acres. The last run of extract was made in 1898. Showing again that Black River lumber was through.

Upon the volume of lumber production during these big years under review the Utica & Black River Rail Road at intervals cast a longing eye. In 1885 the Directors went so far as to initiate a survey to determine the feasibility of running a branch line across the Black River for the purpose of tapping the timber lands of Greig and Watson now monopolized by the canal. Glenfield had the inside track as the probable junction point. Nothing came of the project, a fate reserved for so many U. & B. R. projects.

Nor was this, and the building of the Basselin and Abbey mills, the only source of excitement in 1885. April had seen the highest water reported in the rivers since the disastrous flood of 1869, and under the impact the booms of the G. H. P. Gould mill on Moose River gave way and before repairs could be made a million feet of timber (first reports were of six million feet) had been swept into the Black River. An observer at Glenfield stated that timber there was passing at the rate of a hundred logs a minute. Attempts were made to boom the logs at Bush's Landing and at the Van Amber mill. The booms in both places, however, gave way, but at the new Basselin mill, a half mile further down the booms held and most of the timber was saved.

This was not the only experience of the kind for the Van Amber brothers. In 1877 a million feet of their own logs had broken away and were rescued only when a boom was got across the river some miles below Castorland.

The total of lumber production, as given in the 1879 list, added another 20,000,000 feet estimated for up-river business. At Port Leyden was the J. A. Merwin mill, with 8,000,000 feet, and along the Moose River was Gould, with 7,500,000 feet; Henry Brown, 2,500,000 feet; Comstock and Hyland, 1,000,000 feet, and Russell Barnes and Emory Allen each with 200,000 feet.

Port Leyden was a stirring place in those days, what with the great Merwin plant, the Snyder Tannery, and the thriving Iron Works, taken over in 1880 by interests headed by W. H. H. Gere, a Syracuse engineer who had helped build the canal. Reorganization and enlargement of the plant were on Gere's list of things immediately to be done, and done they were. The village was referring to itself as the "Iron City," and if the Merwin mill was taken rather for granted the fact could be understood, and forgiven. There, however, the mill was, down on the river, fed by an unending stream of logs brought down from the Lyonsdale and Boonville forest-covered hills.

The huge Gould operations had grown out of a number of deals that left him pretty much the king-pin of the Moose River sector, in both the lumber and wood-pulp fields.

The up-river 20,000,000 feet, added to the down-river 28,000,000 feet represented a total of well on to 50,000,000 feet for the canal north from Port Leyden to Carthage. Production made the estimate good for that year, but in subsequent years it soared to new heights.

15

Forestport and Thereabouts

The other center of canal lumbering operations, Forestport, while its tonnage in the '80's and '90's did not compete in volume with the downriver section, yet had men no less picturesque and used methods scarcely less picturesque than could be boasted of by Greig, Watson and New Bremen. And the village, nestling beside a log-covered pond and rollways piled high, made a picture so altogether lovely as one early spring day to move a writer for the Boonville *Herald* to rapturous song.

"Spring," ran the rhapsody, "whose very name brings thoughts of gentle breezes, blue sky and fragrant flowers, is beginning to make its presence felt by that gracious element which it alone commands, and infuses into what otherwise would be a chilly air, making it most acceptable and exhilarating. The hills, having cast aside their wintry garments, now begin to assume a spring-like hue, while down their dark sides trickle little streams of water that but lately was rigid ice or soft snow."

The mood moves on in the declaration that "familiar faces and voices, which have for weeks past been housed in the woods, are now seen and heard among us once again. Those the gentle breath of spring has thawed out and dispatched to their homes, from which they will go forth to engage in the exciting and severe work of driving logs or cribbing the piles and spars, which will commence as soon as spring really opens and the ice breaks up."

The writer might have added that among these faces thawed out by the kindly breath of spring would be many boatmen who, with their tow-teams, had gone into the woods after canal closing the previous fall, would take part in the drives, and, once the mills began running, would again take to their boats.

Elsewhere the reporter had occasion to mention Ed Moran, blacksmith extraordinary, who shod as many canal horses as the next one, and kept as well a livery stable to which canal men turned at such times as the nigh horse might be ailing. Ed held some kind of a record, or so he said, for having shod the oldest horse in Oneida County. The animal was owned by Jonas Puffer and was thirty-five years old. Jack Walsh, of Lyons Falls, who was always shoeing canal horses, admits that Ed was probably entitled to the honors which he claimed. Most of the skeptics at the time, it seems, doubted neither Ed's veracity nor his ability to rise to such an occasion, nor the fact that a horse should live so long, rather they came up with foolish remarks like "Why?" with Grif Evans, Boonville's purveyor of boots and shoes to one and all, replying, when somebody told him about the phenomenal age reached by the Puffer horse, "What did they prop him up with?"

Among the Forestport men were some who seemed never to sleep and therefore had little need of the soothing ministrations of spring—the owners of mills and timber-covered acreage in the hinterland. Of these were men who had started early in the Forestport area and came into the '80's full steam ahead. Men like Phil McGuire, the spar-and-pile king, Denton and Waterbury, Henry Nichols, and Phil Hovey. The activities of these men have already been covered in this narrative, and it is here to present the others who came in during the heavy decades.

Of these one of the more energetic and successful figures was W. R. Stamburg, who came to Forestport as far back as 1851, working in lumber camps and trying his hand at a small mill on Pine Creek, south and east of the village. After nine years of this he returned to Deansboro, whence he had come, and after another nine years returned to Forestport to start upon a remarkable career that saw him owner of the Meeker mill, north of the present railroad station, an establishment described, when he had fixed it up, as "one of the smartest mills in the country." A few years later, to take care of his growing business, he built a mill in the village that became a show place. Especially after steam had replaced water power. The capacity of the village mill was around 80,000 feet a day.

To these two mills Stamburg in time added a grist mill and a general store. His real estate holdings included 4,000 acres of timber, from which he sawed, to feed into the canal traffic as high as 10,000,000 feet a year.

The Forestport mill burned in 1893, an event that brought his lumbering career to an end.

Stamburg used to tell with pride about the time he met Bob Burdette, the celebrated humorist of the *Burlington Hawkeye*, who was coming through the Black River country on a lecture tour. Arrived at Boonville, Burdette was treated to a run over to Forestport on the "Ollie." Skipper Ike Scouten locked into the Forestport pond and took the party, along with a local group that included Stamburg, along the waterfront as far as Woodhull, on the way pointing out the suburb of Toggletown. Burdette rose to the name, and when told about the former names of the village he said he liked Toggletown even better than Punkeyville.

A follow-up of Stamburg's story came in the lecture in Boonville that night, when Burdette announced his continued admiration for the name, and declared that the excursion on the "Ollie" had convinced him that, the Utica and Black River Railroad being what it was, here was the way to travel. The road up from Utica, he said, was "the roughest road I ever shook myself up on. The road bed is probably all right, but the section men leave too many pine stumps and rocks on the track." In one place, he added, the train ran over what he took to be a canal boat. "It wasn't, it was only a saw log."

The next night at a dinner tendered him in Lowville Burdette repeated his observations on the road, and then swung into action with Castorland, a locality of which he had just been made aware. "Castorland!" he declared. "Castorland, only eight miles farther on. What an appropriate place to follow a dinner such as this. Pity it is not an island, then we could call it Castor Isle." Which must go down in literary history as Burdette's, and probably that year's, worst joke.

It should be added, as a footnote to canal history, that the name "Toggletown" was derived, according to legend, from one Eugene Hemstreet, who at the pond-side there repaired canal boats, using, the explanation goes, a toggle in raising them from the pond. In 1885, it must always be regretted, the suburb changed its name to Hillside.

In 1890 the Forestport Lumber Company was formed, and when the sun had set upon the industry it could look back upon a successful cleaning up job. Albany money was behind the concern, with the veteran and popular lumberman, Thomas O'Neil, of Boonville, its manager. The same interests owned the Trenton Falls Lumber Company, and between them the two organizations owned some 90,000 acres of timber. The Forestport company engaged Stamburg to do its sawing. Stamburg thereupon removed to the village the machinery in the old Meeker mill, and so well equipped was he that he was able to turn out five million feet a year for the Forestport Lumber Company and to maintain his own production. He ran night and day to do it. When the mill burned Denton and Waterbury carried on in his place for the Company.

Enos T. Crandall arrived in Forestport from Hawkinsville in 1880, after a few years with a mill at that port, and a previous residence in Boonville. He bought a tract of 325 acres south and east of Forestport, and put up a mill. The cluster of houses that grew up around acquired the name of "Enos," and Crandall established a postoffice there with the same name and with himself as postmaster.

Then there was Joseph Ano, who bought a mill and timber acreage over on Little Woodhull Creek, three miles out of Forestport. Later on he located a mill further upstream.

Ano's lumberjacks were sometimes on the playful side. Back in 1887, when in March he had finished cutting and was busy getting his logs to the mill, the boys took time out to do some coasting on the long hill that sloped to the mill, a quarter of a mile of steepish, in spots sharp turning, logging road. Dragging to the top of the hill the heaviest pair of sleighs to be found in camp, the boys turned back the tongue of the forward sleigh for the purposes of steering, a helmsman was selected, and the craft shoved off. Around curves and pitchholes the craft careened, threatened at each sharp bend by an overturn, and all to the accompaniment, so goes the story, of blowing horns and yelling so ardent that it was heard practically in Woodhull. Unfortunately, they had not worked out a plan for bringing the sleighs to a halt, and were headed straight for a slab pile, but by a quick maneuver, requiring a stout heart and a cool head, they could be shunted at the bottom to another road coming from the left down a slight rise. They made it. Another trip had to be made, and still another, but one journey was all a steersman could make, each at the end of the run having to go to his bunk completely tuckered out.

In 1882 the Forestport combination of English and Traffarn bought a mill at the point where Bear Creek flows into Woodhull Creek—this had been owned by Phil Hovey, who then for a time joined forces with

Stamburg. They were also operating the Stamburg mill at Meekerville, but in course of time it was turned back to Stamburg. Their timber supply was assured, what with twelve thousand acres which they had acquired.

J. H. Wilcox, at various times operating as Wilcox and Smith and as Wilcox and Company, lumbered in the White Lake country, with one of his mills on Long Lake. He had a total capacity of 500,000 feet and better, while Harvey Lewis could saw 600,000 and Jerry App a quarter of a million feet.

Any discussion of canal freighting should list ice as a commodity supplied by the North Country. While a large share of the lumber produced hereabouts was sold in cities along the canal, beginning at Albany and extending as far west as Buffalo, ice went to New York, mostly to the Knickerbocker Ice Company, although in 1890 four hundred tons went to Syracuse from Henry Abbey's mill pond. Forestport cut a good deal of ice, too, the Boyce brothers shipping ten thousand tons and Traffarn and Company an equal amount. Complaints were heard that year about the quality, though not of the quantity, of ice leaving something to be desired. As Joe Boyce said, however, if New Yorkers have got to have their cocktails what the heck! Lyons Falls was in the ice trade, but scorned the canal and shipped by railroad. Some thirty carloads left the village in one April.

The earliest record of ice being taken from the canal for consumption by local communities is found in 1880, when George May, proprietor of the Hulbert House in Boonville, remarked that "from the amount of ice they are taking from the canal basin one would think that Boonville would be in a congealed state during the time that goes here by the name of summer."

In Port Leyden M. Coyle played a careful hand. He first harvested his ice then built an ice-house over it. It was about the same time that the Port was the scene of an unusual accident that could have happened, and probably did, in any canal town. It seems that Maine and Holt one fall had a boat load of corn come in. While it was waiting to be unloaded water got in and presently the corn began to swell. This opened the boat's seams a trifle, which let in more water. This opened the seams still farther, only to let in more water. This, as the story used to be told, swelled the corn all over again, and if the thing had gone on much longer the hull load would have been ruined fer horse feed; as it was it was hurt enough so as Maine and Holt had to sell it at a big discount. A conclusion that was a considerable under-statement, the fact being that the corn was practically given away.

16

Lock Tenders

Lock tenders were of two kinds, Democrats and Republicans. No one back there anyhow ever heard of a Populist or a Greenback keeping a lock. It had to be like that, the way they were given their appointments. The State Engineer and Surveyor, until, in 1878, his functions were taken over by the newly created Department of Public Works, was elected by the people. He selected the Superintendents of the various canal sections, and the Superintendents in turn appointed their lock tenders, all of which meant a turn-over in personnel whenever the political complexion of the State administration changed.

Again, along Section 1 some families were so well represented in the list of lock tenders that it was sometimes said that they were divided into two kinds, the Van Dewalkers and the non-Van Dewalkers. And a Van Dewalker, hearing such a charge, could have replied that no, they were Cronks or non-Cronks, Clarks or non-Clarks, and so on.

Certain of the lock men, it can be reported, were immune to dismissal with State political upheavals. These would be men who merited the distinction by marked ability, and sometimes, as in the case of George Williamson, by having served with distinction in the War of the Rebellion. George enlisted in Company E, 14th Heavy Artillery, coming home minus an arm. Among other qualifications was a full-grown sense of humor. One of the stories that amused the canal was of how one day he swatted a bee that had nipped him a few rods up the tow-path. In an up-surge of whimsy he cut off the head of the creature and left the carcass where it could be found again. Presently a handful of boys arrived at the lock house, a single gun between them, out for big game. A conversation got under way, the lads blowing up their shooting prowess, and George with some whoppers of his own. So much so that the visitors wanted to see some samples of his marksmanship. George was willing. "See that bee up the path yonder?" They couldn't, and George taunted them upon their weak optics and said what they needed was specs. "He's sittin' up there as pretty as you like, just to the left of that mullein stalk." Still they couldn't see the creature. Well I am goin' to blow his head off. Watch!" He aimed, fired, and spurted dust within inches of the bee. The boys dashed up to see what had happened, and there it was, dead as the door nail on Rudy Yost's shanty, "head off clean as a whistle," as the boys told the story back home.

A story was also current of George's endeavors in the debt collecting field. It seems that a lock tender down the canal a piece owed him some money, and all efforts to corral it had failed. He was a duck fancier, was the debtor, and by means of a flock of birds, which were always swimming noisily in the canal, he kept the Sunday dinner a successful institution at his lock house. All other collection stunts failing, George fixed it up with Albert Hoffmeister, a boat hand, that next time through he swing out of the lock and through the bird's hangout drag a line with a hook baited with corn. It worked. A bird bit on the ruse and, when it had been towed out of sight it was brought aboard and delivered at lock 70. This went on until George figured that the debt had been paid in full.

At lock 70 quite a lot of talent was essential. George himself had boated it for a number of years, having owned a couple of boats, and he knew the problems of the boatmen, and the peculiarities of most of them. This alone, however, would not have been enough. The keeper here must also tend the waste-weir that at this point let surplus water out of the summit level and discharged it, tumbling, as pretty as a picture, over a ledge of rocks into Sink Hole Pond. The job called for watchfulness and patience, and Williamson had a good back-log of both. It took more than just a lock and an extrovert boatman to ruffle his poise! There is a story of how, on the evening of July 3, in 1893, a thunderstorm struck, with a bolt of lightning that practically wrecked the receiver on the lockhouse telephone. Asked about it next morning by Byron Scouten, on his way to Utica with a load of Denton and Waterbury wood, George replied, "Oh, that was just me practisin' my fourth of July fireworks."

It is a coincidence that in 1885, George's first year at 70, he had as neighbor at 69 another Williamson, Tom—no kinship so far as these researches have discovered. Then after his retirement in 1895 a cousin, Andrew, likewise a war veteran, came on at 69, later being given 70.

George had as a neighbor, at Lock 60, Rudy Yost, who added much to the more robust humors of the lock tending fraternity. One of the more refined pieces is of how, catching a mess of eels he put them in a barrel of pork to preserve them. To his utter distress he discovered that while it preserved the eels it made them uneatable and completely ruined the pork. And as if that were not enough the barrel itself defied attempts at fumigation. He soaked it in the canal and he scrubbed it with lye, but, as he reported the business, it still stenched. In a moment of desperation he tried to sell it to Ira Hart. Rudy suspected Ira of being the miscreant who from a stance on his boat took a pot shot at his wife, who was picking blackberries just beyond a screen of undergrowth. "Hadn't been that she was bendin' over he'd of got her sure." Had Rudy put the deal over he would have evened the score, but Ira, whom everybody was having grudges against, needed no barrels that day and Rudy in desperation burned it.

Berry picking along the canal was not one of the less hazardous pastimes. The records reveal that the wife of Jim Johnson, of Port Leyden, captain at various times of the "Charter Oak," "Peter J. Rohr," and "Francis S. Norton," narrowly escaped death under similar circumstances.

This same locality served as the setting for one of the more ambitious of Dick Warren's adventures. It seems that Dick set out one morning around about sun-up for a trudge to Boonville. He started so early that lock tenders were yet asleep in their lock-houses and boat people in their snubbed up boats. All except John McAdams with the "R. C. Wade," who, considerate of other people's sleeping habits, was locking himself through 69, presided over by John Stabb. Dick broke in on the scene. "Where's John?"

John nodded towards the lock house. "Dead to the world."

Dick went into immediate action. Without asking for further light he tore up the towpath for Boonville. Here upon arrival he began to spread the ghastly news. John Stabb was dead. Word spread fast and into the farthest precincts of the village. The undertaker was already harnessing his team when relatives of Stabb called on the telephone to assign to him the grisly task of fetching the remains and readying them for burial. The undertaker suddenly bethought him of the coroner, whose authority was important in the case. He had barely routed this official out of bed when he bethought himself of the health officer, and when the three of them finally had themselves rounded up they set out for 69.

Arrived at the lock the party's first reaction was to tantalizing fumes of frying ham, eggs probably nestling beside it, and coffee in the pot. Some kind of wake had been going on probably. The thought quickened their pace, as they burst into the room to be greeted with a cheery good-morning by John, who went about laying the knives and forks. The coroner exploded. "Hey, you, come out and let me get you in the box Doc here brought for you. You're dead."

"Says who who's dead?"

"Dick Warren says who's dead, and he ought to know."

"And you ought to know Dick. See here," and he brandished a fork, "when I shove off I'll be the one to come and tell you and now get the hell out of here."

Such was life on the upper locks when Dick was not down chopping wood for Jay Pixley, or John Stabb was not making whistles out of the young oziers, whittling the ends of them, deft hand with a jackknife that he was, into dainty figurines, a perpetual joy to the kids upon whom he bestowed them.

Not far below 70, living in a house still standing by the highway, was Edwin A. Van Dyke, who, deaf and mute, as was his wife, is remembered by canallers for his skill as a tender and for his unfailing humor. He used to hold that being deaf had its compensations: he couldn't hear any back talk. Nor, he also held, wisely, could he give any.

To the upper reaches of the canal belonged Adam, as he was called along the levels, whose marriage to "Blind Eve" Walter Edmonds built into an unforgettable story. Among the old canal people are those whose reports, when they are asked about the couple, lack the lyrical aspects of the Edmonds account; the others are divided among those who shake their heads and immediately change the subject, and those, mostly the men folks, who will avoid the question by telling about the time Adam and Eve went into a tiff. It seems that during the quarrel Eve threatened to go to Boonville to see a lawyer, upon which Adam helpfully said he would go along to lead her. Locking a boat through a few hours later, he mentioned the business and the captain reminded him that to do that would be to walk right into the arms of the law. Adam said he hadn't thought of that.

Reports of a couple in the Port Leyden sector are of a more jocund nature than stories in the Adam and Eve cycle. In this account, according to Guy Wilcox, Patsy O'Rourke was paying delightful court to Widow O'Reillly, who tended one of the combines of three. And a good keeper she was, too. Before the echoes of a horn or the call of *Rah, lock!* from an approaching boat had died away she had the locks all trimmed and waiting. As for O'Rourke, he was a jovial man, as an Irishman should be.

Patsy was calling on Mrs. O'Reilly of a fair night. He had quaffed just a bit too much. Not much too much, but more than became a man seeking the hand of Mrs. O'Reilly. More than that, Mrs. O'Reilly treated him to a nectar which she had worked up—from some elderberries, it might be.

In good time, and it was a proper time, Patsy made his adieus and made to go home. Outside the door his sense of direction failed him and a moment later he fell into the lock. Mrs. O'Reilly, hearing the noise, ran out and asked was he hurt.

"No, Mrs. O'Reilly, I am not hurt, but it's a dommed high threshhold you have here."

A little more than a mile below lock 70 was another waste-weir, where today the water leaves the old canal bed to pour over rocks in magnificent cascades to join the Lansing Kill, which likewise here comes tumbling into the valley from out of the Potato Hill country. This weir functioned without the attentions which the public paid to the one at lock 70, where picnic parties from Boonville snubbed their row boats and strode across to see Baker's Falls in grand action. When water in the canal was low the falls were no great shakes, but most times a picknicker could, if he were not obnoxious to the lock tender, get the weir gates opened, sending a sudden accession of water thundering over the rocks into Sink Hole Pond. The effect would be particularly startling to a visitor from Syracuse or Camillus standing at the foot of the falls when it occurred, unaware that it was not the elements getting in their dabs but just the lock-tender monkeying with the weir gates.

At 62 the tenders held briefer tenures than those at 70. Some of them could boast of longer service as lock tenders, but not at one lock. Michael Youngs was on the lock in the 1880's, and in various years on 69 and 70.

Eugene, his son, had the lock for a number of years beginning with 1907, and at various times was at 58 and 67. George, another son, was a boatman, owner and captain of the "Elvie and Elmer," named for his two children, and built by Frank Seiter at Hawkinsville. George, who at one time ran the American House, in Boonville, also owned the "Dennis Mulchi," built by Sam Ferguson in Boonville for Lewis Phillips. Phillips had come into the canal country from down Schenectady way, bringing a boat with the engaging name of "Tom and Kit." The "Mulchi" burned in 1894 while being refitted in the feeder for carrying stone for rebuilding the locks at the Five Combines.

Altogether the Youngs family made up a sizable canal-career family. The house in which Eugene lived for many years still stands, opposite the weir, slowly falling into ruins.

Other tenders at 62 were "Veen" Russell, which would be in the late 1870's, and down until Michael Youngs, who in the meantime had been tending lock 70, took hold in 1885. And then came the Joslins. Joseph R. was at 62 in 1895, but before him, as far back as 1881, Sam Joslin had tended 54 and 55. At 61, most of these years, was Francis Joslin, a post that Oscar Joslin had for several years from 1903 on. It can be hoped that Oscar's blind wife, who was never to see the weir-made cascades nearby, could of a quiet evening hear the music of them as they careened into the valley below.

But there are more Joslins to come. In 1896 Eli was bank watch at Forestport. Eli had been captain of the "Clara and Effie," Forestport built, afterwards owned by Ed Ford. In 1910 George Joslin was tending 66, where he was stationed for a number of years at the same time that Jerry and Arthur were stationed at the State shop at Boonville. Nor may Henry, whose services preceded all of these named, be omitted. Henry was captain no less of the "Major Anderson of Fort Sumpter," which has already been locked into and out of these pages. In 1866 he was carrying loads of lumber, charcoal and wood to Whitesboro and Utica.

Only a genealogical survey can work out the degrees of relationship, if in some cases any at all, between the canal Joslins. They came from various sections of the Boonville region. It should be stated, however, that Sam was Oscar's father and came from around Buck's Corners, over across Black River.

Abram Scouten, who tended 52 and 53 as early as 1881, helped to start the name off on a long canal career. The Forestport boating Scoutens have already been noted in these chapters: Isaac, owner of the "Ollie"; Fred, his son; another Abe, Isaac's brother, and Byron his son. After Abram at Locks 52 and 53 came Jay, who in 1885 was tending the Upper Threes (Locks 58, 59 and 60). Thereafter he served continuously at various locks, doing a hitch in the early part of the present century at the Five Combines with his cousin, Roscoe Clark. In 1895 Charley was at 31, and next year Richard, Jay's son, tended 63 and 64. This was the same year that Harvey, who came from around Porter's Corners, was bank watch at Forestport.

Scarcely less numerous were the canal Goldens. In 1885, Michael Golden, who lived at Hillside, tended lock 25—no relation to the Goldens who follow in this narrative. He is best known to history as having a brogue so thick that a knife made no dent in it. Preceding him had been Dave Golden, a great boatman who in good time was to take to the locks. Dave with the "Nellie" cleared at Boonville in 1877 with a load of 40,000 feet of Forestport lumber. The "Nellie" grew infirm on Dave, and later became so leaky that, reports have it, one man was required just to keep the craft afloat. It had been built by Gue at North Western and was owned later in its career by Lafayette Clark and Ed Keyes at Dunn Brook. Other of Dave's boats included the "Georgie Pierce" and the "J. J. Haight," both wild boats—meaning that they were brought in from the outside.

At the turn of the century Dave was tending 50 and 51, and in 1916 he and his son Hayes had the Upper Threes. Another son, Will, who also tended the Upper Three, was another former boatman, at one period owning the "John House," which in 1912 sank between locks 47 and 48.

Of another family of Goldens was Patrick, known up and down the canal as Patsy. Patsy was at 65 practically until the closing of the canal. John, his brother, had tended 63, 64 and 65 as long before as 1891 and on.

17

More About Lock Tenders

Along the exciting Alpine reaches of the canal, roughly from the present Pixley Park to lock 35, the locks were manned, more than by any other family group, by the Clarks. This was the region of the "Gulfers," a term of derision by which the folks on the hills denoted the citizens of "Frog Hollow," this latter in turn a phrase applied to the region by the humorists. On their part the Gulfers got even with the hills by encouraging the legend of the Zekes. This pretty legend was of how in ancient days two brothers lived atop the high hills, one on either side of the Lansing Kill, a good half mile apart. Both were stentorian as to their vocal powers. One of them particularly so, and when of a morning he sought news of his brother Zeke he would face across the valley and give. "H'lo . . . Zeke!" When the ringing welkin had subsided, back would come such data as the occasion required and an inquiry as to the welfare of the other, and what was cooking for the day. For the neighbors a half hour of this was as excellent a medium for picking up useless information as the party telephone line and the radio quiz programs were later to become. And to this day there are wives along the Lansing Kill who, at such moments as their husbands tend to reach up into the higher decible brackets, bring the business to a quick end with a gentle, "Now, Zeke—"

Of all the locks on the canal the Five Combines (the word in local circles accented on the second syllable) were the most exacting, demanding of the keeper an unusual degree of patience, unremitting attention to details and general competence. In any one of the five locks a set of gates could be wrecked by an irresponsible boatman trying to bull in with sixty or seventy tons of lumber aboard. Especially if he was going down stream, with an ever present current to be managed in snubbing into the lock. More than that, the boat would have but a scant six-inch clearance between lock wall and boat, and since the boat practically scraped bottom, the greater part of the water inside the lock chamber would be displaced.

Here was no job for a novice, nor for poorly constructed gates. A misstep and a boat could crash into the lower gates of the upper lock, and, out of control, stop at nothing until it reached bottom, it and its cargo a complete loss. That such accidents were rare may be credited to the skill of the tenders.

Fewer hazards confronted the boatman coming up. At the bottom of the Fives, where the sluiceway, rushing down from the level above, poured into the canal, was a danger spot. Here a steersman, heading for lock 39, swallowed many a cuss word if he was a church goer, and let go with a vivid vocabulary if he was not. A boat must head into the counter currents here set up at just the right angle or might find his boat crashing, with fatal results, against the docking built to guard the lock's approaches. Once inside 39, however, the steersman's worries were over, provided he exercised a decent caution climbing the ascent of the five locks.

The Fives were almost continuously, throughout the canal's operations, in the keeping of members of the Clark family. To a man they had boated, and having run a boat was the best possible education in the intricate business of the Combines. The first of the Clarks had come into the valley from up in the town of Steuben, where on old maps "Clark's Hill" marks the spot.

Roscoe, of all the family, turned in the longest record for service at the locks. He started in 1895 with Abe Hart, and wound up there upon abandonment in 1922. Not that he was at the station continuously: only during Republican administrations. He would have no truck with Democratic canal control, and always upon a switch, even when asked to remain, he went back to boating or other occupations until normalcy was restored.

Two men were assigned to the Combines, a chief tender and a helper. Roscoe's apprenticeship was under Abe Hart, and thereafter at such times as he served he was in charge. Abe and Roscoe got on well together and during one of Roscoe's time-outs they ran the Ferguson-built "Lua and Harry" for G. H. P. Gould, a boat named for Gould's two children.

During Roscoe's long periods of service the second man at the locks might be his step-father, Albert Remp, or his half-brother, Leo Remp. His father, it should be added, Lafayette Clark, had tended lock, though for but a single season when a young man. Until his death in 1876 he was a boatman, among his boats being the "Myrtle Sarah," Gue-built and named for his daughter; the "Willie," a wild boat, and the "Nellie," also made by Gue and named for the builder's daughter.

All this time Roscoe's cousin, Morgan Clark, Jr., an uncle, Orwell, and still another uncle, Martin, and *his* son Martin Jr., known to one and all as "Marty," were for several years in the 1880's at the Fives. The farthest from the Combines the Clarks ever got, as a matter of fact, was, to the south, lock 35, which Orwell Clark tended in the 1880's, until he took over 37 and 38 as his almost exclusive province. Northward the extreme limit of Clark territory was 47, tended for a long term of years in the early part of the present century by Marty Clark. Even Arch Plato, Lafayette's brother-in-law, who with his "Col. J. Van Woert" had been hauling Denton and Waterbury lumber, could not get far from the Clark stamping ground, for in 1895 he took over locks 52 and 53, and in 1900 was at the Fives with Roscoe.

The only boating Clark never to tend lock was Olney Clark, son of Stoars. Stoars, son of Orwell, had seen active service as a captain, among his boats being the "R. C. Wade," and had been a construction boss on the job of repairing the feeder after the great 1897 break at Forestport. He lived at the mouth of the Clark Gorge, near lock 36. His service on the locks ended only with the demise of the canal. Given a little more time Olney like as not would have got into lock keeping, for at the time of abandonment he was yet on the sunny side of his 'twenties.

Further south the Hugenines comprised a small but compact lock-tending family. Peter S. was always keeping locks. At various times he was at 30, 31 and 32 throughout the 1880's and well into the 1890's, ending his career in 1915 at 32. This was the lock on which his brother Jacob started in 1881, and Bert in 1897, Jake having transferred his affections to 52 and 53, and Bert later going, with Chester Nestle, to 33 and 34. The roster closes with Dave, Jacob's son, who, previously a boatman, in 1883 was reported as clearing in Boonville with the "R. H. Roberts," carrying 65,000 feet of spruce lumber bound for Albany. Dave was the family's most active boatman, in addition to the "Roberts" having run the "Charles Downer" and the "Henry Glasser."

MORE ABOUT LOCK TENDERS

To the Hugenine list should be added, though he never boated nor tended lock, Peter F. Hugenine, Peter's nephew, who, an artist living in Rome, threw the entire canal-side into a dither such as the Lansing Kill country never had known. It all happened in 1901, when Peter discovered gold and silver up in the Chase Brook gorge where the stream comes down off Webster Hill into the valley at Dunn Brook. From lock to lock the news spread, especially when the discoverer brought in mining and assaying equipment and found that the rock would yield sixty dollars to the ton, and by this time could report that interested Rome capital was going to back the project. Overnight options all the way from North Western to Boonville, on both sides of the Kill, covered every likely, and even unlikely, acre of land. One old-timer, who could remember the day when news of the gold strike came through from Sutter's Creek, is said to have remarked that the California affair, for sheer excitement, was not a "patchin' to this one right here to home. Who'd thunk it!"

Mining engineers centered their operations at the Chase farm, and among the multitudes Mrs. Chase apparently was the only skeptic: she was too busy meditating upon the board bill that began to mount up to speculate upon the chances of a strike.

From all reports the canal folk kept their excitement within bounds through it all; boatman were used to hearing, in the northern section, of gold being found along the Moose, and silver on other Lewis County streams, and were pretty well insulated against excessive enthusiasm over the liklihood of anything being found in Oneida County. Anyhow it was all over within a twelve-month.

Numerically the greatest of all lock-tending families was made up of the Van Dewalkers, of Hillside. Hillside in those days was a lively spot, boasting of a postoffice bearing the name of Leila (pronounced Leela), a distinction shared by Dunn Brook, though the latter community did not go in for fancy names for its postoffices. Be that as it may, Hillside was the seat of the Van Dewalkers. The lock-tending roll, so far as available records reveal the names, consisted of Squire (there were two of them), at various times at locks 14, 18, 28 and 29; Will, Bert, Isaiah, W. H., Fred, S. B., Isaac, Berlin, Warren, G. H., Martin, George, Clinton, A., and Peter. The author has compiled the list from such sources as are available, but can give no assurance that it is complete; further, he has attempted to reduce the names to a system showing the relationships existing between all of these, and others of the name, but without success. He has sought the expert help of Mrs. Rosella Thornton Turner, a granddaughter of Philip Van Dewalker. Mrs. Turner has contributed no little of Van Dewalker lore to the pages of the Rome *Sentinel*, but declares to the author the impossibility of straightening out the ties which bound one family always to another.

The clan (and it amounted, in the compactness of its interests, to a clan) was identified for the greater part with Cahootus Hollow. The Hollow was a vale which, traversed by a brook coming down off Quaker Hill, flowed into the Mohawk valley a scant half mile above the bridge at Hillside. In this wild, secluded spot lived ten or so Van Dewalker families. How they got there and whence they came remain a mystery. Two Van Dewalkers had farms in the town of Steuben as early as the 'Fifties, but Mrs. Turner is of the opinion that these belonged to another branch of the family. It is agreed that the name was Dutch in its backgrounds, and that at some point in time the clan had come into the Hillside environs from eastern sections of the Mohawk.

Be that as it may, when the town of Western became aware of the Van Dewalker presence they were a numerous colony ensconced in the Hollow. Their topographical backgrounds here enforced a seclusion that did them no good. A simple, honest, credulous folk, they easily found among themselves resources for their amusements and spiritual needs. If now and then one of them journeyed forth into the world outside, an acute home-sickness brought them hurrying home. Of this fact the "Cahootus weed" was a striking symbol. Should an absent one overstay his time the folks back home would say he should have taken more of the Cahootus weed, just as when he set out he was enjoined to make sure he had a good supply of it. One young couple, just married, decided to set up house keeping in a house some two miles distant. They moved in, and by next morning the Cahootus weed had got in its licks. After breakfast they loaded their effects back onto the wagon and came home to the Hollow for keeps.

At one period word began to come through of the success which had come to a Van Dewalker who had migrated to Canada and taken over some land. A proper cooperation between soil, weather and good management had brought him fine crop yields, and some of the younger set at the Hollow decided to follow. They forgot to leave the Cahootus weed behind, and next spring they came trapsing back, one young man walking the entire distance.

Only one Van Dewalker escaped the magic powers of the weed, a young woman who in New York got herself a wealthy husband. In course of time speculation upon how her wealth might be distributed became a favorite, almost a constant, pastime in the Hollow. In the end the only one to cash in on the young woman's strike was Clint Van Dewalker, who sold his chances for two hundred dollars.

Such of the men folks as did not go onto locks worked for "Uncle Billy" Rogers, owner of a considerable farm nearby. Uncle Billy had about him some gruff ways, which, however, covered a mellow heart on the inside of him. Out of some of his gruffer inspirations came the name of the Hollow. Coming upon workers leaning, as the saying goes, on their spades, he had a habit of shouting, "Get on, you Cay-hoots!" And more often than not at the end of the day he would be asking one and another of them if they might be needing a slab of bacon or a few pounds of flour.

Gradually Uncle Billy's phrase came to be transferred by the Hillsiders to the clan as a whole; the clan became "Cahooters," the little valley "Cahootus Hollow," and the brook that babbled by taking its place in history as "Cahootus Creek."

The families have long since dispersed; the automobile had barely reached the region when it began taking one and another of the young folks far afield, as far away even as Rome, fifteen miles to the south. A longing for better ways of living, and an instinct that led them to seek mates from beyond the family circle, led to a breaking up of the community, and now nothing remains, not so much as a road for reaching the spot, now grown over with underbrush . . .

It was a long time since the day when the husband of Maria Van Dewalker nearly fell into the creek in his excitement upon being invited to accompany a neighboring farmer to Rome next morning. Maria was anxious that he leave early so as not to keep the neighbor waiting. She prepared an early supper, unusually hearty against the rigors of the trip, and after supper an early bed. No sooner had the husband fallen asleep it seemed to him than he was awake. Rising, he roused the wife, and after a hearty pre-dawn breakfast he set out for his rendezvous. He was surprised to find the neighbor's house lit up from top to bottom. At the door he was asked what in tarnation brought him over at that ungodly hour. When he explained he discovered that the family had not even gone to bed yet. It was only nine o'clock. The little home in the Hollow not possessing a time-piece the farmer loaned the young man his watch to guard against any more such trips during the night.

18

Still More Lock Tenders

The name of Cronk is represented in boating and lock-tending history by close to a dozen men. Mostly they had come down onto the canal from Ava and from out of the Potato Hill country. They were of varying degrees of relationship, ranking from brothers, and fathers and sons, to, in one or two cases, no near relationship at all. Best known along the canal was Lucian, son of Martin Cronk. Lucian as long ago as the 1870's was tending the Lower Threes. Later, in the 1890's, he was on the State Scow force, while he and Jim Murray had over-all charge, when the canal folded up in 1922, of a group of locks from 71 to 85.

Lucian's most unique contribution to canal history was an ingenious dry-docking expedient that in 1904 he resorted to in preparing his boat, the "Philip G. Hovey," for spring opening. Boonville, where the "Hovey" had been parked for the winter, had no dry-dock facilities. The old Ferguson drydock had been filled in by J. A. Barber, whose growing enterprises needed more and more elbow room. The nearest dry-dock was at North Western, but time was not expendable with Lucian and he found a simple solution to his problem by closing the gates of lock 72 and opening 71; after filling the level all he had to do was to run the "Hovey" in to have as nice a dry-dock as he could wish. And to get room to prevent any sudden attack of claustrophobia, Lucian had simply to close 71 and open 72 to have his boat resting on bottom and ready for calking and other necessary attentions. By opening day the boat was trim and loaded.

Jesse and Ed were Lucian's sons. They boated with him and between times tended lock. Jesse owned and ran the "George F. Weaver," which previous pages have noted as originally built by Frank Seiter for Bill Richardson, and the "Frank E. Gallup," a sand-carrying craft. Ebert ("Bert"), another son, was stationed at 34.

Merrit was Smith Cronk's son, and when not boating he tended locks 37 and 38, some years the pair of them, and other times 37 alone. John was of the Smith Cronk family, and tended 35 and 36. His father, who kept lock 34 in 1895 and 1896, and later on 32, had gone West for a spell, and upon the family's return John married the step-daughter of Orwell Clark, a Schermerhorn, thus linking two of the larger of the canal families. Wesley, of another family of Cronks, came from up Potato Hill way, over near North Pond, today's Echo Lake. In the 1910's he tended lock 51.

A boat, the "Iron Duke," introduced the name of Cronk to canal history when in 1873 it reported the death of its driver, a lad of eleven years—Jimmy Wafful, son of Cornelius Wafful, of the town of Ava. The boat was locking through 72 when he slipped into the lock and was drowned. The boat was listed as being in command of "Captain Cronk," but efforts towards the Captain's identification have so far failed. Nor does the boat show up in available clearance records of the time. Decrepit and all but falling apart, it is reported long afterwards as owned by Austin Rogers, of Hillside. One old boatman recalls that at one time two brothers Cronk, Orrin and Orville, boated on the canal, and it is his opinion that it belonged to one or both of them. What was their relationship to the other Cronks he does not know.

The Nestles, of Dunn Brook, made up a small, compact lock-tending family, headed by Harvey, the father, who off and on (the "on" years determined by the political complexion of the party in power in Albany) presided over locks 37 and 38. This would be in 1896 and 1897, and, later, from 1901 to 1904.

Harvey's sons were Charley, George ("Skip"), Chester ("Chet"), and Sylvester. Charley and Sylvester boated—together they owned and operated the "Jenny Beach," which was built at Bush's Landing by Andrew Beach, and at one time was owned by George Kelley. Charley at another period owned the "John House," built by Roberts in Boonville. Except for a short time when he boated with Charley, Chester tended lock, during practically the same years as his father—in 1895 and 1896 he had 33 and the State dam, down just under the hill from the highway, and again from 1901 to 1904. At North Western was another Nestle, Henry, whose son John, from time to time in the 1890's and down to

1904, was stationed at various times at 19, 23 and 24. Harvey, another son, was a boatman.

The four Pixleys comprised one of the best known of the canal lock-tending families. There were Volsey ("Vol") and Jay, brothers, and Jay's two sons, Jim and Stanley. All were boatmen, and haulers of wood and timber rafts as well, a trade in which they were engaged as early as 1880. In 1883 Jay was on the State Scow with a team, and at intervals tended locks 48 and 62, and in the last days of the canal had oversight of all the locks between, and including, 44 to 66. Jim as early as 1892 was tending 64, and later was at 53. Vol in 1901 and 1903 tended 48 and 49 at Pixley Falls.

The Paddocks, Herbert, Alfred and Willard, constituted a three-brother team, although Willard's chief contribution to the canal was his son, Fred N. Herbert was the only boatman, owning the "Nellie May," which in Boonville in 1883 was cleared with a new captain, Denna Ables, carrying 57,000 feet of hemlock lumber from Forestport, Albany bound. None of the Paddocks tended lock, but were State Shop men. Herbert was captain of the State scow, and later moved to Boonville, where he ran the Central Hotel. Alfred likewise was a State-shop man, in 1914 being boss of the carpenter force. Fred N. from 1902 to 1904 was on the scow force with a team.

The Thrashers were pretty much a family of lock-tenders, and the locale of their efforts was Dunn Brook, where they lived—Dunn Brook and such spots along the canal as might require the services of George and his son George H. Both in the 1880's were on the carpenter force at the State shop in North Western. Wherever locks needed new planking and other repairs, where sagging bridges needed new timbers or lock houses called for tinkering of one kind or another, there the two might be found doing their excellent best. In after years George H. tended 31, a desirable arrangement, since because it was right there that he lived. George H. had two brothers, Frank and Theodore. Frank began looking after locks at 34 in 1895, and the next year went over to 30, a lock that later went to Theodore.

With George at the State shop in 1881 was a man who was already a veteran in the service, Jerry Upright the name. Just when Jerry started upon what became a career these researches have not been able to discover.

He was already settled in the shop in 1873, in that year becoming captain of the new State scow, which had just been turned out in the Ferguson Boonville yard. He continued with the State force until 1895.

Also with George away back there was a figure who opened what for the Hillside country might almost be called the Rogers era. This was Christopher Rogers, whose uncle was the "Uncle Billy" of the Van Dewalker saga. Chris had one of the two teams the State shop was using. He used to tell with a good deal of enthusiasm of the combat between his father, John Rogers, who lived on what is now the Nightingale place just below Dunn Brook, and a person from up in Ava. Some kind of feud between the pair had reached the boiling stage and John, who was of an intensely religious nature, met up one Sunday with his adversary. Words presently were being bandied about, with John declaring over and over again that it was Sunday, and Sunday was not a day for brawling, all in his Dutch rendering of the English tongue, and without benefit of the laws of semantics. The two finally reached the point where action must be the next routine. Suddenly John whipped off his coat, threw it to the ground and, addressing it, said, "Lay there, religion, until I put you back on." Then he went to work on the Ava entry and gave him a clouting that, according to all reports, was a masterpiece, then put religion back on and went his way.

Christopher had a brother Joe, who, however, does not show up as a canal man. Uncle Billy had a son, too—Alfred, who was father to William P. Neither father nor son was ever a locksmith, as Christopher used to call the tenders; they are put in here merely as part of the Rogers record.

Another family of Rogerses in the Hillside and Dunn Brook sections contributed a considerable number of men to the canal effort. Austin has already appeared in these paragraphs as owner of the "Iron Duke." He also owned at one period the "Owego," a wild boat, and in 1903 and 1904 was tending lock 23. His brother, William C., known as "Billy" (giving Hillside its second Billy Rogers), ran the "Emma M. Phelps," likewise a wild boat, and for a number of years until abandonment was keeper, first of lock 25, and then of 26. A nephew William, known, to distinguish him in conversation from his Uncle William, as "Willy," soon after the turn of the century was at lock 29, a few years later being stationed at 30. He had formerly been proprietor of the Hillside House near the village a few miles down. That was a long time ago, so long ago that now in Boonville, he can regard himself as a veteran in still another business, that of dispensing antiques.

At lock 23 in the 1890's was a son of Billy's, Wilbur, who from 1901 on was at 24, while *his* son Earl was on the State wagon, and tended lock 21 during the last two years of the canal's operations.

At Dunn Brook were the lock-tending Races. There were four of them, Henry, the father, and his sons Tom, Fred and George. Only one, Fred, who bought the "Joseph Ano" from Roscoe Clark, boated regularly. On one run in 1892 he had barely got the "Ano," carrying a boat of pulp-wood, out of Forestport when it sank crosswise in the feeder and was successful in holding up traffic for several hours. Henry in the 1880's and early 1890's tended 27, a lock that in 1895 was taken over by George.

Tom Race was at 28 in the early years of the present century, but most of his energies were given to his hotel and grocery store at Dunn Brook. The establishment practically leaned against lock 30—until it burned, that is.

Nearby John Grems also had a store, squeezed in between the canal and highway at lock 31. As if this were not enough, Ed Keyes had a place in Dunn Brook for selling groceries and serving food and potables, whereby it was known far and wide as a café—The Keyes's Café it was called. Ed in 1883 was owner of the "Mina," which he had bought from Reuben Grems. A record of 1876 has the boat as clearing in Boonville with Reuben as captain, and carrying 60,000 feet of hemlock timber from Port Leyden.

Ed had a brother Oliver, who did not tend lock, but was a bank watch in 1896. He owned and ran the "Stuart Keyes," named for his father, which he was running as long ago as 1875, when he was hauling hemlock lumber from Otter Creek. Later, in 1883 it was running with George Wafful as owner and skipper.

And there was Sanford Burch, who in 1881, at 23 and 24, had entered upon a long period of service at various locks, and was to be followed later by others of the family: by Alfred, with whom tending lock 24 was practically a career; by brothers Joe at 19 and 21, and by Arthur at 26. Jim, a cousin, was at 23, and Will at 26.

The Joneses, of the North Western country, made quite a canal name. Luke tended lock over the years, as did Del, and back in the early 1870's Bill was running the "S. M. Ferguson," and later on Gaylord was captain of the "S. O'Connor," rebuilt as the "Mohawk," then having the "J. H. Williams." Mostly, however, the men of the name were on the staff of the State shop. Like Eri, who in 1883 was on the State scow with a team and Jerome, foreman for a score of years, and Glenn, his son. Charley and Tom complete the roster so far as these studies have revealed the names.

Marcus Trask and his son William comprised a two-man tending team. Marcus, who lived at the foot of the Capron dugway, a hundred yards or so above the entrance to Pixley Park, after 1885 was, until 1912, on locks right at his front door, 48 and 49 mostly, ending up at 50. Only once during that entire period was he farther away than 54 and 55. William's first lock came in 1902, when with his father he had 50 and 51. Except these two, his lock career was spent at 49. Neither of the Trasks boated.

The Trasks had as an occasional caller, Bill Marshall, who in the 1910's tended the Upper Threes. Whether he brought along any of his recipes for flapjacks history does not record, but it does state that on one occasion he told of being asked by Charley Pratt in Boonville wouldn't he like to go up and cook for the men in one of the Pratt lumber camps. Bill pondered the offer for a minute or two and finally asked, "Where would be a place for me to eat, Charley?" Charley asked couldn't he eat right there in the shanty and Bill said "Hell no, I couldn't stand eating my cooking."

The Harts mostly took to boating, although Isaiah, father to Ira and brother to Dan, two of the best known Harts, tended lock at 48 and 49 as early as 1880. Dan will be remembered by older canal men as having had a sawmill at Hurlbutville in the same period. Ira was a bank watch, likewise in the 1880's, his watch extending from 33 to 70. Otherwise he was engaged in sundry activities. He had a store at the Five Combines, which burned in 1876. Long before that, in the 1860's, he had owned the "C. E. Stevens," carrying pig-iron from Carthage to Utica. At certain other times he had a store at Hurlbutville and got out wood from along the Lansing Kill. In these and other activities he succeeded in stirring up an unbelievable number of resentments. The nearest he ever came to getting his come-uppances, however, was when Herbert Fitch, a young man from over east of the canal, into one of his columns of news notes which he contributed to the Boonville *Herald* dropped a brief paragraph: "Having built a floating summer residence, Ira Hart and wife are quietly gliding eastward on the raging Erie Canal." The entire valley chuckled, one and all aware that the floating palace was a tow of spars and piles upon which Ira was heading for Troy.

Abe and Tom Hart, brothers, were cousins of Ira's. Abe was associated, as has elsewhere been noted, with the Five Combines. Tom was given the Upper Threes—this Tom is not to be confused with another Tom Hart, a man of affairs who lived in Boonville. Abe's son Walter was a boatman, owner of the "Arthur F. Pilbeam" and the Ferguson-built "George B. Anderson," and ended the roster of the canalling Harts.

19

Disturbing News

In his report for 1900 the State Superintendent of Public Works, John N. Partridge, startled the Black River country by recommending that Section 2 of the canal be abandoned. He presented his case with some statistics that did not make pleasant reading, but that set forth facts that everybody was aware of.

"On the occasion of a tour over the Black River Canal," said the Superintendent, "my attention was attracted to the very small number of boats navigating between Lyons Falls . . . and Boonville. I have since gathered statistics relating to traffic and tonnage on this stretch of canal for the season just ended, and they are presented herewith:

Total number of cargoes northbound from Boonville..........	113
Total number of cargoes southbound from Lyons Falls........	32
Total number of rafts southbound from Lyons Falls..........	2
Total craft	147
Tonnage of northbound boats............................	7,957
Tonnage of southbound boats (including rafts).............	7,703
Total	15,660

"I have been led to compare these figures with those representing the total expenditures for maintenance, operation and repair during the fiscal year of 1900 on this stretch of canal which constitutes Section No. 2. These expenditures were $15,639.74. Thus, strictly speaking, the moving of 15,660 tons of freight between the two terminals of the section cost the State $15,639.74, which is about one dollar per ton. The average rate per ton by boat between Lyons Falls and Boonville is thirty cents. The rate by rail on carload lots is $1.20, and on broken lots $1.60; in other words, the cost to the State per ton for moving freight between Lyons Falls and Boonville, plus the rate charged by boatmen about equals the rail rate."

The subject of abandonment was not a new one. The canal had barely started operations than people in the Legislature or among the higher echelons of canal administrators, were accustomed from time to time to broach the idea of closing the "Little Ditch," along with other canals lateral to the Erie, but let it go at that. This time it was different, partly because John N. Partridge was different. Partridge took office with the quaint notion that one of the duties of the Superintendent of Public Works was to devote a parcel of his time to the Black River Canal. Quainter still was the impulse to look the canal over by means of a personal inspection. It was during his inspection that he became acutely aware of the dilapidated state into which the canal had been allowed to fall.

To make things worse the Legislature was passing through one of its periodical eras of parsimony, and the Superintendent could make no sense out of the meager appropriations made in Albany in relation to the vast amount of repairs to locks and dams necessary if boats were to be kept moving. But facts were facts, and he issued a sharp directive to the effect that superintendents should make no appointments, nor promises of appointments "until they shall hear from me on the subject. It is not possible to say how large a force available funds will make possible. It is desirable that in making appointments consideration be given to men who have given satisfaction, which will encourage faithful service and guarantee better service to the State." In other words a lot of jobs might be lopped off.

In the following year, 1901, Partridge dismissed the superintendent of Section 2, John F. Jones, for not following orders and demanding bills on the delivery of goods and supplies and forwarding such bills in due form to Albany on or before the fifth of the following month. Charles E. Searles, of New Bremen, was appointed to occupy the vacant post.

Furthermore, in this year Partridge returned to the subject of abandonment, repeating his stand of the previous year. Section 2 was fast becoming resigned to its fate, when the man whom it regarded as its nemesis was succeeded, beginning with 1902, by Charles S. Boyd. Any uplift of spirit which news of the change inspired along Section 2 was dissipated when Boyd also took up the cudgels for abandonment and gave some new figures, covering 1901 and 1902:

	1901	1902
Cargoes northbound from Boonville	123	90
Cargoes southbound from Lyons Falls	8	3
Tonnage, northbound boats	10,710	7,940
Tonnage, southbound boats	964	360

Naturally this could not go on for long, but when the night was blackest rumors spread along the canal that the State was projecting an extension of the canal to Watertown. Hopes began to rise, and received a shot in the arm when it was announced that a new Superintendent would follow Boyd in 1905. The change was indeed made, N. N. V. Franchot advancing to the post. Franchot, however, picked up where his predecessors left off, and in his first annual report to the Legislature declared that traffic "north from Boonville, which for a number of years has been so small as not to justify its continued operation, has fallen even lower than in previous years, the total number of cargoes north from this point [Boonville] being 41 and the return cargoes being limited to five, with a total burden of only 4,048 tons." And elsewhere he declared that in view of the conditions with regard to business north of Boonville "it seems clear to me that it is a duty resting upon the legislature to provide for the abandonment of the Black River Canal north of Boonville, and I most earnestly recommend that this be done."

Franchot went as far as he could go in cutting operating costs by consolidating the section for administrative purposes with Section 1.

Franchot's statement was, however, in the way of an anticlimax, for it had been preceded that year by the message of Governor Higgins, who went all-out for the abandonment of Section 2. "The Constitution of the State (Article VII, section 8) prohibits," he informed the Legislature, "the sale of the Erie Canal, the Oswego Canal, the Champlain Canal, the Cayuga and Seneca Canal, or the Black River Canal, and imposes upon the State the perpetual management of the same. Whenever it appears that any portion of the canal system has so far survived its usefulness as to make its maintenance by the State a burden, with no corresponding benefits, the Legislature should submit to the people the

DISTURBING NEWS 137

proper constitutional amendment to permit the abandonment of such portion."

The only result of all this pressure was nothing. All canal traffic between Lyons Falls and the river country to the north ceased from "natural causes," a phrase much in use in canal circles to cover all deterioration due to simple neglect by the State. The section was fast abandoning itself, and from around 1905 no lockages into the river, or from the river into the canal, occurred. The author has attempted to ascertain the boat that was last to pass through lock 108, but no two reports are of the same craft and even if the matter were of any particular moment it is unlikely that the secret will ever be out.

And so came to an end one of the most glamorous chapters in the story of the canal—Lewis County timber, which had seemed endless in supply, was gone; areas in Watson, Greig and New Bremen that had fed hemlock, pine and spruce timber into scores of gang saw mills for half a century were deserted, and the old canal men were left to their memories —a roll of men that would include William Clobridge and Tom Kingsbury, who tended the Whittlesey bridge, and W. I. Stillman and John Studor at Beach's. And Merrick Chapman and M. Ervin at Glendale, George Barker and Robert H. Barnes at Castorland, with William Parker and Chris Warner at Castorland. And in the list would be Ezra and John Merry, who were at the Bush's Landing lock and Clinton Tiffany and J. H. Clobridge at Otter Creek. And men who kept the channels open and otherwise made navigation of the river possible—men like A. C. Wilder and George Jackson, Henry Fenton and John Linedecker and John Kirley. And Henry Wetmore, who long had been one of the best known of the Black River boatmen. The river portion of the canal looked back upon its past; the reaches from Boonville to Lyons Falls had not been counted out by the Legislature, but the future looked dark enough, when in 1911 came an event that put the latter back into business for a short while, and another event that brought to Section 2, from Lyons Falls to Carthage, a glimpse of maybe a new life.

The first event was the start toward construction of the Delta dam. The contractors had to have stone, thousands of boat loads of it, and they turned to the old friend of all canal contractors, the Sugar River quarries. Every boat that could stay afloat was brought into service; boats that creaked at the joints for old age were patched up—about the only boat, in fact, that didn't get a try at hauling stone was Benjamin F. Terry's "Lottie Corser." Benjamin, better known up and down the canal as "Joe Torrence" and as the "Human Horse," had found the "Corser" in the Boonville basin, sunk, and old beyond any help, so they said. All except Joe. Joe patched the derelict and somehow got it afloat, made himself a harness and began towing the boat up and down the canal, hauling wood to points as far away as Utica. Bringing up a load, bucking the current, was too much even for a frame as huge as Joe's, so his per-mile profit would have put any other boatman in the red, what with horses and crew to feed and with other overhead items that a canal man had to worry about.

Obviously the "Corser" did not go into the stone business. Neither, for that matter, did "Muskrat" Haley's little steamer. This was a boat made by Muskrat himself. It was a side-wheeler, propelled by an upright engine that all but laid the hull wide open when the going got tough. Obviously, here was no craft for the stone trade. It was Muskrat's home; he lived abroad, staying in ports like Boonville until a desire for new sights took him on to North Western and other places. When in course of time the outfit got as far away as New London the boat, engine and all, gave up the ghost.

The Delta dam would have been more of a god-send had it been larger; as it was a few months saw the reservoir completed. By this time, however, occurred the second of the two events. The excitements supplied by this one were provided by the Legislature, which voted an appropriation to survey a route whereby the Black River Canal would be extended from Carthage to Sackets Harbor, necessitating the reconditioning of the Lyons Falls to Carthage portion. A number of surveys were made, but the plan favored by most engineers was, according to John A. Bensel, State Engineer, the canallization of the river from Carthage "to within a short distance of Deferiet. Then comes a cut-off to the east of this village, till the river is regained a little to the north. From this point to the eastern end of Huntington Island, just east of Watertown, the river is used, with the exception of a cut-off across the bend at Black River village. The river channel is used around the north side of Huntington Island and then a land route is followed which extends along the northern outskirts of Watertown, passing, near its western end, into Cowan's Creek and entering Black River at the mouth of this creek, a point which is opposite the fair grounds. Then the river is again utilized to Glen Park, where a line to the south of the river is begun, which runs to the headquarters of Muscalonge Creek, and then follows this stream to its entrance into Lake Ontario through Muscalonge Bay, which is an indentation some three miles from Sackets Harbor." The estimated cost was $16,300,000, the prism and locks to be equal in size to those of the Barge Canal.

The next year, which would be 1912, a bill was actually introduced in the Legislature authorizing $16,000,000 for construction. As introduced the bill outlined a project quite as simple as that. It was too simple for Albany processes at such times as canals and railroads were under consideration; as finally passed the project included also the rebuilding of the old Chemung Canal and the improvement of the Glens Falls feeder on the Champlain Canal. The bill got through both houses of the Legislature but was vetoed by the Governor.

All of these doings should have raised the temperature of the down-river population, but there is no evidence that more than an occasional eye-brow was raised. The region had seen the 1911 Legislature vote $50,000 for rehabilitation of the canal between Boonville and Carthage, only to have the appropriation all used up when the work northward from Boonville got only as far as Lyons Falls—and when a bill authorizing another $50,000 to finish the job to Carthage was introduced and had passed both houses the Governor vetoed it!

The public was prepared, therefore, for the next move of the Legislature, which in 1913 revived the extension project, and went still further in complicating the chances of acceptance by the Governor. Not only did it include the Chemung and Glens Falls projects, but added the rehabilitation of the Delaware and Hudson canal, running from Rondout to the Delaware River, and the construction of a proposed canal to connect Flushing Bay on Long Island with Jamaica Bay. Again the Governor vetoed, giving as his reason the fact that none of the projects save the Black River extension had been adequately surveyed. The Assembly the next year voted money for all of the surveys demanded by the Governor, and when these were completed it appropriated the sum of $68,000,000 for going ahead with the projects under consideration—all except the rehabilitation of the Delaware and Hudson job. The bill passed the Assembly of 1914, but failed in the Senate.

20

The Pulmotor Fails

From the collapse of the Sackets Harbor extension project nearly four years elapsed, when the rapidly disappearing canal traffic convinced the Legislature that the day of decision had about arrived. Choice must be made between total abandonment and a restoration of the entire waterway. Annual tonnage had dropped from 143,561 tons in 1889 (the all-time high save for 1910, when the canal was carrying Sugar River stone for construction of the Delta reservoir) to 71,447 tons in 1897, 58,013 in 1907, 47,262 in 1914, when the Sackets Harbor business had been dropped, and 14,821 in 1917. Moreover, of these 14,821 tons more than nine thousand tons were made up of Boonville sand, a commodity that, a natural for a railroad haul, might be expected any day to leave the canal. Goods that had been the backbone of carriage in the old days had all but disappeared from the canal. A further breakdown of the 14,821 tons shows, in addition to the 9,702 tons of sand, 2,114 tons of wood, 443 of potatoes, 1,002 of coal, and 1,560 of general merchandise.

Potatoes, which long had been a staple item of canal business, was to drop from the 443 tons of 1917 to 225 tons in 1918, and, beginning with 1919, played so small a part in the year's cargo lists as not to be given as a separate listing. A brief review of potato carriage may be given as an illustration of the steady deterioration of traffic on the Black River Canal over the previous forty years and more. And the available figures first strike the reader with the slowness with which potato production developed in the canal regions, as compared with that of lumber, for example. Mostly the early reports include, without identification, potatoes in the general list of agricultural products. It is known, however, that in 1857 7,062 bushels were shipped, with only 3,564, however, in 1867, ten years later. In 1876, however, potatoes arrive at the distinction of a separate classification under agricultural products, with 70,573 bushels, which in 1879 became 90,517 bushels, and 152,232 bushels in 1880. This latter was an all-time top figure and subsequent reports reflect an up and down production. The following figures are for years that, with years of fluctuating tonnage between them, yet show the steady direction of the downward curve: 72,600 bushels in 1895; 52,033 in 1900; 25,410 in 1903; 20,464 bushels in 1916; 4,942 in 1918 —and that is the last report in which the potato is given a listing of its own.

The potato traffic added notes of color of its own to canal lore. There are old boatmen today whose most pleasant memories are of going, in late autumn, with father to New York with a boatload of potatoes to sell, finding a berth probably in the Wallabout Basin, in Brooklyn, where the potatoes would be sold out of the boat, to dealers and consumers alike. Three thousand bushels made up a standard load, and selling three thousand bushels usually was a winter's job. The team was left behind in one of the Troy's, at stables which let the horses out for their keep to farmers in the vicinity, although some boatmen brought, or sent, their horses back home.

A few months later, with spring come and the potatoes gone, the party, aboard their boat, came back up the Hudson, towed as part of a fleet of other boats, recovered their horses and went on home.

On such occasions as the load was sold out early in winter the party returned home and the skipper of the boat went to New York prior to canal-opening time, with ice gone from the canal, to bring the boat home. Henry Wetmore, who hauled potatoes from the Watson country, recalls that he sold potatoes, not only in Wallabout but at the foot of Grand Street in New York. Another boatman tells with gusto of the winter when, with half of his potatoes sold, he was waited upon by some minor functionaries, who informed him that he would have to get the heck out of there, which was Wallabout, and into another berth. This could have meant almost any one of a number of things: such as moving over into New York. Or into one of the Morristown Canal basins over in Jersey City. Or finding a way to stay where he was. A visit to a well-placed party in New York brought about an amicable adjustment and he was able to stay put in Wallabout, to the consternation of the local gentlemen.

Selling potatoes, even in Wallabout market, was not a particularly inspiriting way to spend a winter. It had the virtue of being profitable, however, but just the same when spring came and time was near for canal opening, it was all right with the owner of the boat and such of his family as were with him. Charley Boyce, of Forestport, in 1885 traded his horses, wagons and sets of sleighs for Mike Johnson's boat, the "G. H. P. Gould," and that fall loaded up with a cargo of Phil McGuire's potatoes and headed for Wallabout. He seemed anxious to get there, but on returning in the spring he said that Forestport looked good to him after a Brooklyn winter. And should a boatman respond to any lure that Wallabout possessed, by getting home quickly after canal opening he could load up with stored potatoes and carry them down for spring business.

Such instances were not too frequent, for many of the boatmen would have spring chores to do around the little place which was their home, and if a boatman was a farmer, there would be crops to be planted and a look around for profitable summer business. In any case, however, a man would be come October putting a load of potatoes aboard and setting out for Wallabout.

Potatoes at the time of their harvest were a cause of many a headache to all concerned. If the crop was light prices would be high, which pleased only those who were fortunate enough to have at least a fair yield, but for the buyer there would be too small a volume of sales to make a profitable season of it, while for the boatmen cargoes would be too few. On the other hand, too big a yield pleased only the boatmen; prices would be low and at such times the farmers always threatened to let the crop rot in the ground, but in the long run they sold; the buyer like as not stored part of them for the winter and shipped them down in the spring, but autumn or spring, a haul was a haul. As when in 1878 thirty boat loads were waiting in the Boonville basin for the canal season to open. That was the time when Ike Scouten loaded his boat, the "Fred and Wilber" and slipped out of the basin ahead of opening day, with three thousand bushels of potatoes aboard being shipped by Jacobs and Traffarn, buyers. The same buyers the following autumn shipped out of Boonville three boat loads, A. H. Barber six cargoes, and Henry Glasser thirteen loads. Hawkinsville was a considerable potato mart and shipping point, and the spectacle of three and four boats being loaded there at one time was not an uncommon sight—one of the regulars there being the "Esther and Jennie." The operations of these buyers, along with Thompson and Company, likewise of Boonville, extended to down-river areas, like Port Leyden, their favorites being such varieties as the Rose, Hebron, Burbank and Red Chili. In this down-river area, however, they found competition in the activities of Homer Markham, of Lyons Falls, one of all the largest of the operators.

Prices during a buying period fluctuated sharply, and with a considerable range, requiring alertness on the part of buyer and producer alike. At thirty cents, which was a kind of standard low, the farmers were given to holding back—as in 1886, when Forestport reported that only two boats were being loaded, and these by Charley and Joe Boyce. Thirty-five cents would bring potatoes out of hiding—sometimes set up too late, as when in 1863 B. P. Miller, of Lowville, opened buying with forty-two cents. That was more like it, the farmers apparently thought, for they reacted to the price so heartily that later, in response to some law of supply and demand, he reduced his price to thirty-five cents, only to find that the deluge of potatoes was practically over. After all of his trials, however, he was able to ship six cargoes that fall.

There were years, of course, when a bumper yield sold as low as twenty-five cents; farmers, however, were accustomed to holding back their potatoes for anticipated price advances. Sometimes they were successful, and sometimes they were not.

The last few years of the canal's operations saw new men in the field. Among these were Tom Shanks, of Forestport, and Walter Furman, buying in Forestport and Hawkinsville—wherever they could find potatoes, for that matter. Walter himself, on his farm over beyond the Black River from Boonville, was raising three thousand bushels a year. Another producer whose crop, either all or in part, went to Shanks was Sam Dorn, down off Jackson Hill, known in those parts as the "Potato King." Sam was producing six thousand bushels a year, and when not selling to Shanks might be selling to Will Grems. Will and his brother Mike had made up one of the canal's most popular brother teams. Will had held an important post in the construction of the Delta, having in charge all of the boats and barges bringing stone, sand and other materials used in the works. He had owned the boats "Florence," named for his daughter, and the "John E. Mason." The team of Shanks and Furman owned three boats, the "Walter Furman," the "Harrison and Kenneth," later sold to Charley Moyer, and the "Paragon Plaster Company"—all lakers built by Doran in Durhamville. Later on they had the "Minnie E. Matton," a Troy-built boat. In addition to these, boats were leased from Jesse Cronk, Warren Tuttle and others.

Great as had been the business of raising and hauling potatoes, however, it seemed all but over when the Legislature in 1918 voted $170,000 to put the canal back into operating condition; it seemed over, just as lumber had gone out thirty years before, and as everything else was going, except sand, which anyhow could not provide the volume to maintain a canal in the way that in more prosperous times it had been accustomed to.

Once the appropriation of 1918 became law the Department of Public Works went to work with a will, losing no time in putting the money to work. The job was put in the capable hands of the late Wilkes Dodge, then of Boonville, who from 1895 had been associated with practically all of the Department's repair operations on the Black River Canal. In his hands plans were quickly drawn up; overnight contracts were executed, and men, materials and machinery were assembled, as if to get the work completed before the Legislature had time to change its mind. The work was actually got under way in June, 1918, and by the end of the season navigation had been opened as far as the half-way bridge between Port Leyden and Lyons Falls. In addition, the river locks at Otter Creek, in the town of Greig, and at Bush's Landing, in the town of Watson, were being rebuilt, as also some of the more dilapidated works at Lyons Falls. It did not augur well for the project that the Bush's Landing and Otter Creek locks alone would consume $30,000 of the appropriation.

In 1819 work was resumed with the same vigor as before, and at the end of the season the Superintendent of Public Works was able to report that dredging the channel between Lyons Falls and Carthage had made such headway that "it is expected that navigation on the waterway will be established in midsummer [of 1920] as far north as Carthage, the point specified in the act. The placing of the canal again in a condition for use will undoubtedly bring to the Barge Canal system a considerable increase of commerce. Of the $170,000 appropriated there is an unexpected balance of $40,000. This must be reappropriated by the 1920 Legislature for continuation of the work."

THE PULMOTOR FAILS 145

It cannot be reported that the boatmen along the river went into a fever of optimism and experienced any excess of excitement. The very enthusiasm and efficiency with which the contractors pursued their work aroused doubts in the minds of the old-timers, particularly at such times as politicians went into transports of elation over the way things were going. They had seen too many starts made at rehabilitation of the canal, only to see things revert to what was a normal state of neglect and dilapidation when appropriations made by the Legislature had been used up. Thus it was that when it was revealed that only $40,000 of the appropriation remained to finish the job, and that during 1919 $82,467.41 of the $170,000 had been used, any ardor that may have been lurking among them would have been confined to the smallest possible number of cases.

Reporting progress for 1920 the Department of Public Works could report that "the work of improving the Black River Canal, under authority of Chapter 564, laws of 1918, was continued during the year 1920. To make possible the use of locks reconstructed during the preceding year, dredging operations were had in the river sections between lock No. 102, located south of Lyons Falls, and . . . Whittlesey's Bridge, and from Glenfield north to and beyond the Otter Creek lock.

"No dredging plant having been available in the vicinity, suitable units were constructed by department forces. In the building of these units the necessity of transferring them elsewhere was had in mind. With this in view, one of the dredging plants was comprised of two hulls, each fourteen by seventy-five feet in dimension, and when lashed together constituted a hull 28 feet wide. The other outfit used consisted of a hull fourteen by seventy-five feet, which was equipped with two eight-foot pontoons on either side. With the completion of the improvement these outfits will prove valuable for use at other localities on the canal system."

And the report goes on to say that "in addition to excavating material from the river channel, considerable length of wall was built through the village of Lyons Falls, and bank protection placed in the vicinity of Otter Creek and Beech's [Bush's Landing] lock. It is expected that early in the coming season of navigation the entire project will have been completed, thus making the Black River Canal navigable as far north as

Carthage, which was the purpose of the original act." The cost of the year's operations as described above was $32,349.03.

Other work in the meantime had been going on, under the supervision of the regular canal authorities, and paid for out of the normal annual appropriations for ordinary repairs. Farm bridges were being put in repair along the southern section, and a few new ones built. New lock gates were being installed where necessary, and sluiceways and weirs conditioned against the new flow of traffic that the new tomorrow would bring. The department performed a prodigious amount of work, considering the miserly appropriations that were placed at its disposal. The stark fact remained, however, that at the end of 1920 less than $10,000 of the $17,000 rehabilitation appropriation remained, a fact that all too clearly portended the end. The Superintendent in his report for 1921 was bafflingly reticent upon the project; no word about what was, or what was not, done during the year; no new appropriations were mentioned. No word of regret, nothing of explanation. Only silence. And some of the old boatmen smiled.

21

Dreams Don't Jell

The Black River Canal could never make its dreams come true like the Erie had a knack of doing. All the Erie had to do was think about bigness for an hour or two and next day Mr. Expansion himself would be knocking at the door, bringing men and equipment for letting out the prism here and there, and making longer and ever longer locks.

The Little Ditch just didn't have the touch. It never asked for much, which maybe is where it erred. Its chief hope was that some day it might be kept efficient. It already had the most beautiful route of any of the State's, perhaps America's, canals. It wanted more than that—it wanted to reach a state where all of its locks would be working, if possible at one and the same time, and to have lock walls that did not threaten constantly to cave in on the boats or to disappear in sink holes. And it wanted tow- and heel-paths that did not wash out at every smart freshet that struck down off the hills, ripping out two and three hundred feet of bank and laying up boats and their cargoes for weeks on end. The maintenance forces did their best under the meager appropriations given them to work with; at long intervals they could even get around to painting the lock houses and keeping the tow-path clear of weeds and brush.

At one now forgotten period the canal ventured to dream of enlargement. There were no existing or foreseeable conditions of traffic that would warrant enlargement actually, but the same could be said, if you wanted to get railroady about it, for the Erie Canal. The little one soon had a sizable agitation under way along the Black River levels. By the time the movement had reached the stage of action, however, and a call was sent out for a convention to be held in Boonville on January 19, 1887, "enlargement" had been cut down to mean merely deepening the canal to the extent that boats drawing four feet of water could be navigated. To this proposal was added a call for the completion of the upper reservoir at Forestport, which had been authorized in 1883—and, in spite of the convention's presentation, was not to be completed until 1893.

To the meeting came the greats of the entire region. There were James Galvin, Superintendent of Section 2 of the canal, and John L. Norton and Charles Ryther and C. J. Clark, L. H. Mills and R. Rickeman, all of Carthage; the Hon. G. H. P. Gould, William S. DeCamp, Marshall Potter and Walter Whittlesey, of Lyons Falls; Charles Denslow, of Port Leyden, and Watson Van Amber, of New Bremen. From Alder Creek came the veteran canal surveyor, able and beloved by all who knew him, Chandley L. Phelps, and from Forestport a large delegation that included W. R. Stamburg, Alonzo Denton, F. X. Salzman, Isaac Scouten, James Boyce, Philip McGuire, Henry Gallagher and John Root. Locust Grove was represented by the Hon. Clinton Levi Merriam, Congressman; North Western by boat builder Jerome V. Gue, and Hawkinsville by S. O'Connor. T. B. Basselin, of Croghan, would have been on hand, but he missed a train in Utica and sent a telegram expressing his hearty support of the movement. Active in the proceedings of the afternoon were Boonville's Peter Phillips, Superintendent of Section 1, Lyman W. Fiske and T. S. Jones, attorneys, and former State Senator Robert H. Roberts—the latter brought the meeting to order and presented Carthage's John L. Norton to be chairman of the proceedings. Frank Willard, of Boonville, and DeCamp were named secretaries.

One of the most useful speeches was made by Norton upon assuming the gavel. It is given here in full: "Gentlemen—I thank you heartily for the honor and partiality shown in choosing me chairman of this meeting. What is your further pleasure?"

A speech equally to the point was made by Surveyor Phelps. Disappointed that the meeting was not to go down the line for enlargement, and was willing to settle for a mere deepening of the canal, he devoted his speech to the observation that the canal in its first days did accommodate a draught of four feet, and that all that was needed now was to clear the bottom of silt and other accumulations of the years to have their four feet, and what are we waiting for?

All the speakers agreed that $40,000 would be enough for the job, and the difficulty of getting this much money out of the Legislature probably justified the length of the resolutions, which consisted of three whereases and one long resolved. This one resolution provided, among other things, for the appointment of a committee to wait upon the Legislature and by some kind of wizardry wangle the money out of it.

The best speech of the day was that of J. L. Clark, who, speaking as a Carthage shipper, declared that the prosperity of all northern New York depended upon the canal. Without any question the canal had, more than any other single factor, been instrumental in building up the region's industries. The production of talc, iron, mica and other minerals was being developed, new railroads were being built, as witness the new Carthage & Adirondack line, "the pride of every Carthage heart." Every year a million tons of ore would be brought, mostly by the C. & A., to canal-side in Carthage. And all of this tonnage must be carried southward by canal. As of that moment the Utica & Black River Railroad was carrying it because of the insufficient draught of the canal.

It was the concensus of opinion among the delegates that a Black River canal association should be formed by shippers and boating people, a kind of pressure group that could exert mass influence in places where it would count. The suggestion did not get into the resolution and it then and there ended its brief existence. Which was just as well—the plea for $40,000 had a good spot in the resolution, but little good it did. Nothing ever came of the business, and dreams of greatness henceforth were left to the larger and gaudier canals, like the Erie and the Oswego.

The canal did turn aside now and then to consider better forms of motive power than that supplied by horses. Steam was such a medium, and the proponents of steam power were soon to have an efficient example, plying chiefly between Forestport and Boonville — the little "Ollie," owned by Ike Scouten and Horace Dayton. Steam, however, required that the skipper have a pilot's license; Ike and Horace were the only pilots between Forestport and Rome and between Lyons Falls and Rome, and the two of them alone could not plug with any great effectiveness for their favorite motive power. Bob Dayton, Horace's son, later on piloted the "Ollie," when his father had bought Ike's interest in the boat and changed its name to "Helen" for a grand-daughter. Bob lays claim to having been the youngest pilot ever to run a boat on the canal—he got it at fifteen. The force of pilots was thus increased by a third, but it had no visible effect on the universal lack of interest in steam.

"Snubbing Posts" likes best to meditate upon a novel means of canal motive power thought up by Tom Shanks and Walter Furman when one day they found themselves behind in their schedule, with a boat load of potatoes lying over in Hawkinsville that had to be moved quickly. Horses weren't fast enough and the answer of the hour was to hitch the tow-line to the back end of their new Chalmers, crank the car and hope that all the tires were on straight. They need not have worried; everything went off without a hitch, and a half hour after shoving off in Hawkinsville they were snubbing up in Boonville. Tom, asked for the precise time, said, however, it was so little as not to be worth mentioning. That was in 1914, in the later days of the canal, when everybody had become reconciled to the canal's ultimate abandonment, and nobody is on record as trying to emulate the exploit and so help to get up the Chalmers as standard towing power.

The most glittering idea for a substitute for horses in that dream age was of a narrow-gauge railroad, to be built on the tow-path. This contrivance, providing of course that it would work at all, had one good point: it would operate summer and winter, in winter hauling freight up and down the canal in cars suited to the traffic. What would prevent the little railroad from running its cars also in the summer and putting the boatmen out of business was never divulged.

This one, like the dream about the trolley system, soon passed. If it served any useful purpose at all it must have been to help keep conversation going during such dull days in the North Western and Hillside sections as no chickorees were being reported, or further north boatmen were sweating out a tie-up brought on by a break.

A chickoree, it may be explained, was a nocturnal pastime indulged in by the more prowlingly inclined of a community's youth. The main idea was a raid on somebody's hen-house. From the hen-house the raiders took their booty to a convenient cove, or other remote spot affording adequate cover, and proceeded to barbecue it. Should the owner of the roost, who in more cases than not was a lock tender, be heard to mutter about it next day he might be waited upon by one of the delegation and asked, with an innocent air, just what, if he were confronted by one of the raiders, he would do about it. The question might be accompanied by a demonstration of a man driving a fist practically through a plank, or other convenient object short of a boulder. Thereupon the offended party was almost sure to make it clear that nothing whatever came to his mind.

Oxen likewise were advanced as a likely substitute for horse-power. This thought, strangely enough, had some support, and by lovers of horses and oxen (few as the latter were) alike. The former seemed to feel that pulling a canal boat was no proper work for a horse, while the latter felt that at the business end of a "hooser" an ox would be in his element. The discussion never reached the agitation stage.

Discussion of the electric trolley system came up the canal from the Erie, as did most other ideas having to do with motive power. The plan in most of its details followed rather closely the operation of the electric street car. In the stern of the boat would be placed a dynamo of from fifteen to twenty horse-power. This contrivance would operate a stern screw, similar to that of a steamboat. Power would be drawn from a wire strung above the canal; or, to quote an engineer's description current at the time, "Along the side of the canal at appropriate places will be poles, and across the water will be strung a wire to support the trolley wires." The trolley pole would have free lateral movement and, as worked out by the promoters, several boats could thus use the same wire. Power was to be got into the trolley wire from power stations stationed at intervals of twenty miles. A pleasant thing about the plan was that boats would move at a speed of four miles an hour instead of the two-mile speed provided by horses.

The trolley enthusiasts got nowhere at all on the Little Ditch—probably because there were not enough of them. One suspects that most boatmen were in the same state of mind as Roscoe Clark, when somebody asked him what he thought of the idea, and he said that it wasn't "hell, rain or shine to him." Or of Jim Ross, a lock-tender from over Point Rock way, who asked how, for a boat pulled by a dingus like that, he'd have to "harness" the lock—an expression, equivalent to "fitting" a lock, that so far as these researches have been able to make out was Point Rock's only contribution to canal lore.

It cannot be urged that periodical attempts by well intentioned people to close the canal on Sunday were on the list of its dreams. The promoters of the idea regarded the spiritual improvement of the canal as no whit less important than the improvement of the river between Lyons Falls and Carthage and consequently their efforts may be given notice in this place. It all went back to the early Erie days, when periodical attempts were made to get the Legislature to make a law about it. One year, back in the '40's, the Legislature took note of a flood of petitions from all over the State; a committee was appointed in the Assembly, and the committee in due time reported. In presenting some of its observations the committee remarked that "it may be safely received as a postulate, that every effect partakes of the precise nature of the cause; therefore, any evil intent, whether it be manifested in breaking the Sabbath, or in any other evil doings, or whether it be manifested at all, pollutes the spiritual atmosphere, and by the laws of sympathy, operates through this medium upon the minds of the whole community, more or less prejudicially; thereby making every member of the human family directly interested in banishing evil thoughts from the minds of all mankind. We frequently see this interest manifested by those who know not why they feel it, but are impelled by it to suffer all that may be suffered in this world, with but a faint hope of banishing a few evil thoughts from their fellow-creatures."

When "Snubbing Posts" read all this to Clarence Davis, down in Western, the old canaller's answer was "Ribbety chuck!" Which may be loosely translated as Western's equivalent of "phooey." And this was a common sentiment along the canal at such intervals as the agitation for Sunday closing came up.

Such few good breaks as the canal ever got were without the benefit of dreams, referring particularly to the building of the Delta reservoir. Excepting that it gave a lot of work to boatmen and their boats, it would not be listed by many of the told-time boatmen as necessarily a dream job. But the reservoir did make accessible to the Erie Canal a vast quantity of water, and provide it almost immediately as needed; it was pretty to look at, and the owners of land appropriated by the State were well enough compensated. The village of Delta, of course, lies at the bottom of the pond; farms and farmsteads are no more. Locks 7 to 13 inclusive, along with the old viaduct between locks 8 and 9, were inundated, part of them, and the others abandoned. From near lock 13 the canal had to be relocated, following closely the present State road, and dropping down to river level by means of a three-lock combine still to be seen at the dam-side.

For the Erie it was a splendid achievement. If any of the more nostalgic of the old boatmen were inclined to mourn the change they could accept it in good enough part—they were perfectly aware that the end of the canal was not far away.

22

Curtains

The wonder is that construction of the Delta reservoir could bring into service the many Black River boats for the hauling of sand and stone that it did. From 1901 on so many boats went into hiding, or were abandoned, that shippers, in such seasons as potatoes went into big crop figures, or a sudden demand for stone or sand appeared, went into mild states of despair. Moreover, canal officials in Albany were expressing concern and calling for measures that would entice boatmen back to the levels. Not the slightest attention, however, was paid to their complaints; boat shortages on the Erie canal had been chronic for years, and, by some strange logic used in Albany, enlargement in the form of the Barge Canal was going to bring new boats, whole fleets of them. That was the Erie, however, and while enlargement for the Black River was discussed, and urged throughout the valley, and renovation was actually started later on, neither course would have been any more successful in re-boating the Black River than enlargement was able to do for the Erie.

Thus it is that the student of the Black River canal is surprised to discover how many of the old-time boats came out of hiding for the Delta enterprise, in addition to the thirty-odd barges which the State caused to be built for the job. He is even more surprised to discover that many of those that responded were veterans of thirty and forty years of service and that some of them were able to navigate at all. The author is indebted to Ralph Shanks, son of the late Will Shanks, old Forestport boatman, for a list of Black River boats made by his father. From this list, and the boats that have appeared in previous chapters, came the boats that helped build the Delta dam. Boats like the "Robbie," which was Will's own last boat, and the "Abram Cox," which had belonged to Alex Phillips, and at one time to Brom Willis, of Port Leyden; George Thorp's

"Alonzo Denton," and Henry Coscomb's "Allen Phillips." Even the "Maid of Judah," Bill Roe's old boat, was in the list; incredible as it seems, this redoubtable craft had been navigating the canal since 1866 and evidence exists which seems to show clearly that long prior to the Delta works the "Maid" was completely worn out and had been discarded, and possibly destroyed.

Then there is the "G. R. Ainsworth," which had been owned, at different periods, by Charley Boyce, of Forestport, by Will Griffith, of the town of Western, and by Allen Phillips when he was living at the Five Combines. And the "C. W. Eaton," a Henry Randall boat that later had belonged to John Cannon, and the "C. W. Colton," which Ike Scouten had owned prior to buying the "Ollie." The "Colton" sometime later was known as the "Charley and Donald," at a time when it was owned by Jim Donovan, of Forestport, named by him for his two boys. The "Clara and Effie" is another ancient one that belonged to Ed Ford, of the same place—a record of as long ago as 1883 shows it as clearing at Boonville and captained by Eli Joslin.

Then there is Will Dorrity's boat, the "Charles Dorrity." The Dorrities were of North Western; Charley had been Superintendent of Section 1, and gave a performance that well deserved the honor of seeing a boat named for him. A boat, it may be observed, might bear the name of the owner's wife, a child, or a pair of children, or of a friend. Or a boatman might name the boat for himself, as in the case of John Donley and the "John Donley." All this, it must be conceded, was better than handicapping a boat with a name like the "City of Persepolis" or such. A canal man could, and sometimes did, attach to his boat the name of some worthy citizen, or even of a corporation, should either be desirous of parting with a smallish sum in consideration of having his or its name spread along the Black River and other canal routes. It was a pleasant custom and, since the amount which changed hands was not deductible from anything, no base motives can be imputed to either of the parties involved in the transaction.

The "Charley," of North Western, built by Gue and owned by Ezra Sherman, of Boonville, was another veteran—it cleared in Boonville as early as 1876 and, as Ed Lake, who used to tend lock 70 and later was on the State scow at Boonville, used to say, it was old enough to be President of the United States. The "D. E. Dillenbeck" (Douglas Dillenbeck was a North Western merchant) was built in the Gue yards by Jim Davis, of Dunn Brook, who was a great hauler of potatoes. Later it was owned by Will Roe, of Westernville, husband of the energetic Emma, who has previously appeared in these pages.

Also listed is Forestport's own "Electa," owned, built, named and run by Albert Harrig. And the "Emory Allen," best remembered for its associations with Henry Brockert, of Boonville, who had owned it over long years, having bought it from Lon Denton, of Forestport. It was an old boat at the time, Mrs. Brockert informs the author, for in addition to Lon previous owners had been Harvey Boyce, who cleared with it at Boonville as long ago as 1875, and Charley Davis, brother of Mrs. Brockert. Henry afterwards owned the "J. A. Petrie" and the "Maggie Cullen," a wild boat, which he bought from Mike Johnson, of Forestport and later sold in order to buy the "Petrie." The last of the Brockert fleet was the "Hart and Crouse," which is also in the Shanks list, bought from George Ward at Rome. He had already abandoned the "Emory Allen" because of sundry infirmities of age.

George Root's "Elmer E. Nichols" is on the list, and Sidney Phillips' "Dudley Capron," and Jim Gallagher's "Freddie and Harold," Jim Gawkins' "Frank and Ollie," and Warren Tuttle's "Gold Lock," the latter run at one time by Brom Willis. The "Gen. J. A. Hill" was there, too, a veteran of thirty years' service—John McCann had owned and run the "General" in the early 1880's. And Bill Griffith's "Henry Patton," which Ed Van Schaick had also owned; Gene Benedict's "H. B. Stevenson," hailing from Whitesboro; Harry Ford's "Henry Stephon"—Harry was a son of Ed Ford, who had owned the "Clara and Effie" and the "Thomas O'Neil," and a veteran in service, the "B. F. Cady," which Roscoe L. Cady had built for him by Gue, and which he named for his father, Ben Cady. It was at one time run by Walt Wood.

Forestport was represented by still another boat, the "Henry Nichols," owned by the Nichols brothers, who were deep in lumber, Dave Barry its skipper, and by Mike Donovan's "Jennie and Lizzie," and by Jim Boyce's "Lumber Boy," and by the "Martha," owned variously by George Warren, Albert Harrig and George Weaver. Also of Forestport were the "Phillip G. Hovey," which had run under a number of skippers—Robert Thorp, for one, and Jess Cronk, and Gus Agley; and the "W. R. Stamburg," Barney Oaks captain, and Will Morreall's "S. M.

Ferguson." Morreall, while hailing from Forestport, yet lived impartially for a time in Port Leyden, where he had a livery stable, and in Boonville, where he was proprietor of the Park Hotel.

Then there is the "W. O. Leland," owned by one Trandell. Trandell, not a native of these parts, had come to the Black River from Hinsdale with an ancient craft, long overdue at the junk yards, named the "General Grant." This he sold to Stillman Quackenbush and soon was running the "Maggie Cullen."

On the list, too, is the "John and Jeanette," Christopher Leaf its captain, previously owned by Andrew Belknap, of Port Leyden. The "John and Jeanette" at the time of the Delta proceedings must long previously have reached the age of superannuation for boats, and consequently is here put down on the discarded list alongside the "Maid of Judah" and the "Emory Allen." And there was Will Grems's "John E. Mason," and another veteran, the "James Hyland," of Boonville—the "Hyland" as long ago as 1873 had cleared in Boonville with Sam Sage as its skipper, and in 1883 with Charley Curran. Another Boonville boat was Nick Cannon's "Leander W. Fiske," built by George Seiter, a sister boat of the "Main and Holt," and still another, the "Schweinsberg," owned by Nick Schweinsberg, Boonville shipper, its captain in the early days being George Boardman, and later on Charley Thrasher and John Shanks, older brother of Tom and Will.

North Western contributes the "Lura C. Gillette," owned by Lura's father, Horace ("Hod") Gillette, and Port Leyden the "M. W. Holt," Dell Satterly captain, while the town of Greig adds the "Sarah Jane," owned by George H. Plato. The "Oscar Gorman," Bill Richardson, closes the list.

Upon completion of the Delta reservoir the boats went right back into hiding. By 1916 only 410 boat loads were cleared in Boonville, for a total season's tonnage of 20,464 tons, or a little under fifty tons a boat. Five years later, in 1921, only eighty-four boats checked out, for a total of 4,225 tons. There is no report of clearances for 1922, the last year of the canal's operation; the tonnage, however, was reported—150 tons, which would mean, with normal loads, but three clearances for the entire year. And all three loads are reported as being made up of "products of the ground," which was official language for sand.

For 1923 the Superintendent of Public Works says laconically in his report, "There was no transportation on the Black River Canal during the navigation season just ended."

That was it! Just like that. No official notice of the canal's abandonment. As a matter of fact, the canal never was officially closed. So far as any action by the canal authorities went, the canal is in official existence today equally with its status in 1922. The difference is that a boat trying to achieve the trip from Boonville to Rome today would find the journey even more hazardous than it had come to be in the closing years of navigation. And that, old boatmen would claim, would be saying a great deal.

What boat was last to go through has by now become a matter of conjecture, and even of mild argument. Clarence Davis, of Dunn Brook, enters a claim for Sid Phillips and his "Dudley Capron." Sid was a son of Allen Phillips, and nephew of Pete and Alex, and the claim, could it be corroborated, would be one more for the record of a distinguished boating family. "Snubbing Posts," however, has found no evidence upon which a decision could be made with any clear substantiating evidence, and is willing to leave the matter open, one subject for old boatmen, so long as they shall be around, to argue about. On all others they are agreed: they are agreed that it was a grand little canal. That it was great fun while it lasted. That it had the worst locks to be found this side of Kingdom Come, and the best lock tenders. That it had less water, and on what there was ran more freight per gallon than any other canal in the whole blamed universe. That it had the best fighters, and less, along the Black River canal levels, to fight about than all the other State's canals put together.

One more boat was to go through: the Boonville State scow. That was let down to Rome in 1924—the precise date July 24. Phil McGuire was, or had been the last, Superintendent and looked after the procedure. He and the hands with him must have wondered if the trip was worth while; the task meant filling the old prism from water turned in at the waste-weir below lock 63, and doing some tinkering on a few of the locks, which already were showing the effects of neglect and inaction. But the scow got through without mishap, and the Little Ditch called it a day, and rolled over and went to a long sleep.

Postscript

The Black River Canal from first to last was in all of its ways uniquely a North Country, and therefore a homey, kind of institution. Its habits, its ways of doing things, its customs, the tales which grew up around it—all of these came out of the nature of the region which manned its locks and boats, and which supplied the freight that it carried. As illustrated by the predominance of horses over mules on its tow-path.

The great majority of the boatmen worked winters in the woods, hauling logs to such mills as were located in the early times, up and down the streams, and in later days to such streams as floated the timber down to the mills that came to be concentrated on the canal at places like Forestport, and on the lower river at Basselin's, and Van Ambers' and Abbeyville.

The mule was considered as not suited by nature to logging and was looked down upon as a creature wayward, without any of the traits of human kindness. The general attitude toward the creature was summed up in something a writer did for the Boonville *Herald,* who observed that "when spring comes and navigation on the canal is resumed, the mule returns to his work without a change of countenance or any unusual excitment. The winter season is his holiday time, but he does not spend it in the primrose paths of dalliance nor the flowery beds of ease. The relations between the mule and his driver are never cordial, but there is mutual respect, regulated on the mule's part somewhat by the driver's command of language. If the mule should be talked to as you and I talked to the horse he would treat you with undisguised contempt. He doesn't bear any malice against his driver, but when the latter inadvertently gets between him and the canal the mule does not hesitate to kick him over the berme bank and will scarcely take the trouble to look around to see whether he has rolled back into the water and drowned. His principal characteristics are his voice, his rock-ribbed sides, his sense of humor, his patient toil, his practical immorality, and his ability to live and work almost without food."

Tom Shanks was the one boatman that had a preference for mules over horses for the tow-path, while from older boatmen one hears now and then, sketchily, of a white mule, always, however, so faint a figure as to raise it practically to the status of a phantom towing mule, the first mule possibly ever to have achieved the distinction.

The character of the canal as the product of a locality was further heightened by its tendency from the first for boating and other canal activities to shake down into family groups. At any moment a lock-tender might be letting through a brother, or a cousin of first, second or third rank, and uncles of the same grades, or in-laws without number. And if it wasn't a relative it would be somebody who knew a relative. All of which contributed to the development of a folksy character that could not prevail down on the Erie, where hominess paid homage to hurry.

And the finest touch was given by the many wives, with their children, who accompanied their boating husbands, keeping the boat tidy, the larder filled, and providing as much as could be the comforts of home, even on occasion the bearing of children. Among these was Jess Cronk, born on his father's boat in lock 70; it is an interesting fact that his nephew, Erv Lake, was born in the lock house at 70, his father tender there at the time. John Brockert, son of Henry, himself to become a boatman, was another boat-born child.

And there were occasions when with her husband a boating wife spent the winter aboard, the boat tied up, and winterized so far as could be, at a convenient spot. Mrs. Betty Hugenine, who boated for many years with her husband, David Hugenine, informs the author that one time two winters in a row were spent by them tied up in the Boonville basin, finding the winter-bound craft a most pleasant and comfortable abode.

Out of conditions like these grew, the student of the canal discovers, those qualities that made the Black River unique in many ways among canals, and that provided memories that today make pleasant dreaming for all who lived and worked along its picturesque levels.

www.ingramcontent.com/pod-product-compliance
Lightning Source LLC
Chambersburg PA
CBHW032039290426
44110CB00012B/874